RAND NATIONAL DEFENSE RESEARCH INSTITUTE

Review of the Provision of Job Placement Assistance and Related Employment Services to Members of the Reserve Components

Agnes Gereben Schaefer, Neil Brian Carey, Lindsay Daugherty, Ian P. Cook, Spencer Case

Prepared for the Office of the Secretary of Defense

For more information on this publication, visit www.rand.org/t/rr1188

Library of Congress Cataloging-in-Publication Data

ISBN: 978-0-8330-9177-2

Preface

The Office of the Secretary of Defense asked RAND to conduct a congressionally mandated study, which, in Section 583 of the 2015 National Defense Authorization Act, requires that "the Secretary of Defense . . . conduct a review of the feasibility of improving the efforts of the Department of Defense to provide job placement assistance and related employment services directly to members in the National Guard and Reserves." The aim of this study was to identify the current universe of federal employment programs and resources available to reserve component members, assess the gaps and overlaps in those programs and resources, and develop recommendations for how the U.S. Department of Defense (DoD) and the federal government as a whole can improve efforts to provide job placement assistance and related employment assistance to reserve component members. This study consisted of three tasks: (1) review the literature on federal job placement assistance and related employment services for reserve component members, (2) conduct informational discussions with managers of federal programs designed to provide job placement assistance and related employment services to reserve component members, and (3) assess the feasibility of improving DoD efforts to provide job placement assistance and related employment services to reserve component members. The study was carried out over the course of 11 weeks, from March 11, 2015, to May 27, 2015.

This research should be of interest to federal policymakers, managers of federal employment programs, and others concerned with how to improve employment assistance to members of the reserve components.

This research was sponsored by the Office of the Assistant Secretary of Defense for Reserve Affairs and conducted within the Forces and Resources Policy Center of RAND's National Defense Research Institute, a federally funded research and development center sponsored by the Office of the Secretary of Defense, the Joint Staff, the Unified Combatant Commands, the Department of the Navy, the Marine Corps, the defense agencies, and the defense Intelligence Community. For more information on the RAND Forces and Resources Policy Center, see http://www.rand.org/nsrd/ndri/centers/frp or contact the director (contact information is provided on the web page).

Contents

Tables

Summary

The Office of the Secretary of Defense asked RAND to conduct a congressionally mandated study, which, in Section 583 of the 2015 National Defense Authorization Act, requires that "the Secretary of Defense . . . conduct a review of the feasibility of improving the efforts of the Department of Defense to provide job placement assistance and related employment services directly to members in the National Guard and Reserves."[1] The aim of this study was to identify the current universe of federal employment programs and resources available to reserve component members, assess the gaps and overlaps in those programs and resources, and develop recommendations for how the U.S. Department of Defense (DoD) and the federal government as a whole can improve efforts to provide job placement assistance and related employment services to reserve component members. This study consisted of three tasks: (1) review the literature on federal job placement assistance and related employment services for reserve component members, (2) conduct informational discussions with managers of federal programs designed to provide job placement assistance and related employment services to reserve component members, and (3) assess the feasibility of improving DoD efforts to provide job placement assistance and related employment services to reserve component members. The study was carried out over the course of 11 weeks, from March 11, 2015, to May 27, 2015.

Since the study team could not find a compilation of all federal employment programs that provide employment services to reserve component members, and since our research sponsor was not aware of such a compilation, our study approach consisted of identifying the universe of employment programs and services available to reserve component members, identifying potential gaps and overlaps in those programs and services, and identifying potential ways to improve employment services to reserve component members. Our findings from tasks 1 and 2 resulted in the development of a compendium of federal programs, resources, and offices that provide employment services to reserve component members. Our findings from tasks 1 and 2 also informed task 3 and our effort to identify gaps and overlaps in federal employment programs, resources, and offices that are available to reserve component members. Finally, our findings from all of the tasks also informed our recommendations for how to improve federal efforts to provide job placement assistance and related employment services to reserve component members.

---

[1] Public Law 113-291, Carl Levin and Howard P. "Buck" Mckeon National Defense Authorization Act for Fiscal Year 2015, Section 583, December 19, 2014.

Methodology

To document the federal programs, resources, and offices that provide job placement assistance and related employment services to reserve component members, we used a systematic and varied set of approaches. We started by conducting a broad literature review of employment programs available to reserve component members and then collected as much information as we could from program websites and other publicly available material. In addition, through our conversations with key stakeholders from each of the major agencies that provided job assistance to reserve component members, we received some recommendations for programs to include and some additional information on employment programs for reserve component members. These searches were guided by a data-abstraction form that allowed us to document information systematically across programs.

One of the goals of this study was to identify potential gaps and overlaps in the portfolio of federal job placement assistance and related employment programs that provide services to reserve component members. We focused on two sets of key questions to guide our analysis:

1. Do all members of the reserve components have access to some type of employment assistance? Are some receiving more assistance than others? Are there gaps or areas of duplication?
2. Are there perceptions among key stakeholders about other ways that employment programs can be streamlined or expanded to better serve the needs of reserve component members?

We used several methods to explore these potential gaps and overlaps. Our first step, as detailed in Chapter Two, was to account for the key characteristics of the federal programs, resources, and offices that provide job placement assistance and related employment services to reserve component members. For each program, resource, and office, we identified the types of services provided, the intensity of those services, and the populations that are served. We then assessed this summary information to determine where gaps and overlaps might exist.

Our second approach was to conduct informational discussions with key stakeholders from each of the major agencies that provided job assistance to reserve component members. These key stakeholders shared with us basic information about programs, provided feedback on our full list of programs to identify gaps in our data, and shared some perspectives on where gaps and overlaps might exist and how they might be addressed.

Since the study team could not find a comprehensive needs assessment that identified the employment needs of reserve component members, and because our sponsor was not aware of such a needs assessment, to assess potential gaps in federal employment programs, the study team developed a framework in which we assessed the programs' services against the needs that reserve component members have expressed in previous work we have conducted on this topic (e.g., the need for assistance with legal issues), as well as typical employment-related services for the general public.[2] Based on our review of programs and our discussions with stakehold-

---

[2] For perspectives on reserve component employment needs, see Laura Werber, Agnes Gereben Schaefer, Karen Chan Osilla, Elizabeth Wilke, Anny Wong, Joshua Breslau, and Karin E. Kitchens, *Support for the 21st Century Reserve Force: Insights on Facilitating Successful Reintegration for Citizen Warriors and Their Families*, Santa Monica, Calif.: RAND Corporation, RR-206-OSD, 2013. Also see Laura Werber, Jennie W. Wenger, Agnes Gereben Schaefer, Lindsay Daugherty, and Mollie Rudnick, *An Assessment of Fiscal Year 2013 Beyond Yellow Ribbon Programs*, Santa Monica, Calif.: RAND Cor-

ers, we identified several areas of potential gaps and overlaps in federal programs that provide employment assistance to reserve component members.

Caveats and Limitations

The timeline for this study shaped the scope and methodological approach in important ways. First, the findings are limited in their scope to the program data that were publicly available. As a means to double-check the accuracy of the information we found during our literature review, we originally planned to conduct informational discussions with the managers of all of the federal employment programs we identified. However, given DoD information-collection constraints under DoD Instruction 8910.01, *Information Collection and Reporting*,[3] we were not able to do so. If the research team had conducted more than nine informational discussions with DoD personnel, we would have had to acquire a report control symbol (RCS) license. However, given the very short timeline for this study, it was not feasible to acquire an RCS license; therefore, the research team conducted only nine informational discussions.

Additionally, a complete analysis of gaps in services or an evaluation of the adequacy in employment assistance would have required a comprehensive assessment of the employment needs of reserve component members. Given the absence of such a needs assessment, we instead identify potential gaps by assessing federal programs against employment needs that have been articulated by reserve component members in prior research, as well as common employment-related services for the general public. As a result, our findings regarding potential gaps in employment services are quite limited, and they do not take into account the full range of employment needs among the reserve components.

In addition, our current analysis describes the supply of programs, but it cannot compare these programs with demand for employment services or their adequacy in addressing the needs of reserve component members. While we identify areas of overlap and areas where fewer services are provided and point decisionmakers to specific areas of concern, our analysis is not sufficient to determine whether existing programs are meeting reserve component members' needs and which programs are the most effective and efficient at doing so. To determine which programs are best meeting the needs of reserve component members, full program evaluations would be necessary.

It is also important to note that our analysis of gaps and overlaps focuses only on federal programs. State programs and nongovernmental organizations also offer a range of employment services for reserve component members that might help to fill gaps and add additional areas of overlap.

Lastly, while we were able to identify most programs' eligible populations, program services, and the intensity of those services, we were not able to evaluate the effectiveness of those activities or the programs in general. As indicated in our recommendations in Chapter Four, a formal evaluation of the effectiveness of federal employment programs would be a critical step

---

poration, RR-965-OSD, 2015. The study team also identified typical employment services offered by private employment agencies and state employment offices.

[3] DoD Instruction 8910.01, *Information Collection and Reporting*, Washington, D.C.: U.S. Department of Defense, May 19, 2014.

in identifying how DoD and the federal government as a whole could improve job placement assistance and related employment services to reserve component members.

Therefore, this study represents a preliminary assessment of this important topic, and there are opportunities to expand and enhance the analysis in the future. For instance, with a longer timeline, it would be feasible to acquire an RCS license and conduct more interviews with directors of federal programs that employ National Guard and Reserve members. In addition, a comprehensive needs assessment would enable a complete assessment of gaps in services or an evaluation of the adequacy in employment assistance for reserve component members.

Summary of Findings

In total, we found 40 federal programs, resources, and offices that provide job placement assistance that can be accessed by reserve component members. The bulk of the job placement assistance and related employment services available to reserve component members are provided by agencies in DoD, Department of Labor (DoL), and the Department of Veterans Affairs (VA). It is important to note that many of the programs (including most of the VA programs) are targeted to a limited population (e.g., veterans or service-disabled veterans), to which only a small portion of the reserve components belongs.

Potential Gaps and Overlaps in Services Provided

We first identified employment needs that reserve component members articulated in previous research, as well as common employment services provided to the general public to meet general employment needs. These categories of services include assistance with résumé and interview preparation; assistance with job search and placement; assistance with job and career planning; skills assessment and certification; job fairs; assistance with legal issues; assistance with entrepreneurship; internships, apprenticeships, and training; financial assistance; and access to information and tools. We then accounted for whether a program, resource, or office offered each of these ten categories of services. The information was compiled into a matrix that allowed us to visually identify gaps and overlaps (see Table S.1). In addition to accounting for the services provided by each program, resource, or office, we also accounted for the intensity of services. We define *high-touch* programs as those that provide opportunities for extended follow-on services that are typically personalized to meet individual needs; we define *low-touch* programs and resources as those that provide one-time services for a short period, that do not provide personalized services, and those that do not provide any human component. Of the 40 programs that we identified, we categorized 20 of them as high-touch programs.

For the most part, we did not identify any gaps in services in the current portfolio of federal job placement assistance and related employment services available to reserve component members. As indicated in Table S.1, the most commonly offered services include access to information and tools, assistance with job search and placement, assistance with job and career planning, and assistance with résumé and interview preparation.

While we did find that a broad set of services is covered by existing programs, and that the stakeholders we interviewed did not perceive any significant gaps among the many federal programs and resources that are available, we also found some potential areas of overlap. There were a large number of high-touch programs providing basic employment services, and while many of these programs target different populations and offer unique approaches to service

Table S.1
Services Provided and Populations Served by Federal Job Placement Assistance Programs and Resources

Agency	Program Name	Population Served	Targeted to Reserve Component	Assistance with Résumé and Interview Prep	Assistance with Job Search and Placement	Assistance with Job and Career Planning	Skills Assessment and Certification	Hosting Job Fairs	Assistance with Legal Issues	Assistance with Entrepreneurship	Assistance with Internships, Apprenticeships, and Training	Financial Assistance	Access to Information and Tools
DoD	Airman and Family Readiness	Air Force RC members	No	●	●								●
	Always a Soldier (AAS)	RC members with service-related injuries	No								●		
	Beyond Yellow Ribbon (BYR) Programs	All RC members	Yes	●	●	●		●	●				●
	Career Skills Program (CSP)	Activated Army RC members	No		●	●	●				●	●	●
	Credentialing Opportunities Online (COOL)	All RC members	No			●	●				●	●	●
	DoD Hiring Heroes	RC members with service-related injuries	No	●	●			●					●
	Education and Employment Initiative	RC members with service-related injuries	No			●	●						●
	Employment Readiness Program (ERP)	Army RC members	No	●	●	●	●	●		●			●
	Fleet and Family Support Centers	Navy RC members	No	●	●	●		●		●			●
	Guard Apprenticeship Program Initiative (GAPI)	Army RC members	Yes								●		
	Hero2Hired (H2H)	All RC members	Yes	●	●			●					●
	Job Connection Education Program (JCEP)	Army RC members in four states	Yes	●	●								●

Table S.1—Continued

Agency	Program Name	Population Served	Targeted to Reserve Component	Assistance with Résumé and Interview Prep	Assistance with Job Search and Placement	Assistance with Job and Career Planning	Skills Assessment and Certification	Hosting Job Fairs	Assistance with Legal Issues	Assistance with Entrepreneurship	Assistance with Internships, Apprenticeships, and Training	Financial Assistance	Access to Information and Tools
DoD	Marine and Family Program's Career Services	Marine RC members	No	■	■	■							■
	Military OneSource	All RC members	No										■
	National Guard Employment Network (NGEN)	All RC members	Yes	■	■	■	■						■
	Operation Warfighter (OWF)	RC members with service-related injuries	No			■					■		
	Partnership for Youth Success (PaYS)	Army RC members	No		■								
	Public Private Partnership (P3) Program	Army RC members	Yes	■	■						■		■
	SkillBridge	Transitioning RC members	No				■						
	Troops to Teachers (TTT)	Transitioning RC members	No						■				
	Yellow Ribbon Reintegration Program (YRRP)	Activated RC members	Yes	■			■	■		■	■		■
DoL	American Job Centers (AJCs)	All RC members; veterans receive additional services	No	■	■	■							■
	Career OneStop	All RC members	No		■								■
	O*NET and My Next Move—For Veterans	All RC members	No	■	■		■						■

Table S.1—Continued

Agency	Program Name	Population Served	Targeted to Reserve Component	Assistance with Résumé and Interview Prep	Assistance with Job Search and Placement	Assistance with Job and Career Planning	Skills Assessment and Certification	Hosting Job Fairs	Assistance with Legal Issues	Assistance with Entrepreneurship	Assistance with Internships, Apprenticeships, and Training	Financial Assistance	Access to Information and Tools
DoL	**Registered Apprenticeship**	All RC members	No		X						X		X
	Veterans' Employment and Training Service (VETS)	All RC members	No	X		X						X	
VA	**Compensated Work Therapy (CWT)**	RC members with service-related injuries	No	X	X		X				X		
	Education and Career Counseling and CareerScope	Transitioning RC members; RC members qualifying for educational benefits	No			X	X				X		
	Office of Small and Disadvantaged Business Utilization (OSDBU)	Veteran RC members	No							X			X
	VA for Vets	Veteran RC members	No										X
	Veterans Employment Center (VEC)	All RC members	No		X				X				X
	Veteran's Opportunity to Work (VOW)	Veteran RC members completing VocRehab	No										X
	VetSuccess on Campus (VSOC)	Veteran RC members; RC members qualifying for educational benefits	No		X		X				X	X	
	Vocational Rehabilitation and Employment (VR&E)	RC members with service-related injuries	No		X		X						

Table S.1—Continued

Agency	Program Name	Population Served	Targeted to Reserve Component	Assistance with Résumé and Interview Prep	Assistance with Job Search and Placement	Assistance with Job and Career Planning	Skills Assessment and Certification	Hosting Job Fairs	Assistance with Legal Issues	Assistance with Entrepreneurship	Assistance with Internships, Apprenticeships, and Training	Financial Assistance	Access to Information and Tools
VA	Workstudy Program	RC members qualifying for educational benefits	No		■							■	
Other federal agencies and joint programs	Feds Hire Vets	Veteran RC members	No						■		■		■
	National Resource Directory	All RC members	No										■
	Transition Assistance Program (TAP)	Activated RC members	No	■		■				■			■
	Veterans Business Outreach Program (VBOP)	All RC members	No	■						■	■		■
	Veterans Employment Program Offices (VEPOs)	Veteran RC members	No	■	■								■
Number of programs, resources, and offices offering services				16	23	15	11	6	4	6	13	6	29

NOTES: Programs in bold are ones that provide high-touch services. RC = reserve component.

provision, these programs could potentially represent overlap. In addition, there are many programs that offer slight variations on very similar tools and resources, and these resources could potentially be streamlined to cut costs and ensure consistency in information. Finally, while outside the scope of the study, overlaps with state, local, and nongovernmental programs must also be taken into account when planning and implementing federal programs.

Potential Areas of Concern

Our discussions with key stakeholders suggest there are some potential areas of concern if programs are consolidated. Based on these discussions and prior studies on employment services for veterans and service members, federal agencies should consider maintaining the following types of programs to avoid future gaps in services: (1) high-touch programs, (2) programs that have strong connections to employers, and (3) programs that have strong data-collection and case-management capabilities.

In addition, given the large number of federal employment programs and resources offered by different agencies and designed to serve different populations, a major finding that emerged from our search is that it is extremely difficult to navigate the large number of programs and resources available to service members. Other studies of programs for service members argue that overlap and a lack of coordination can make navigation difficult.[4] It was even difficult for our research team to tease out the relationships and differences between some programs and resources. When we presented our list of programs and resources to one of the stakeholders that we spoke with for his feedback, he commented that if he were a service member, he would not know where to start if he were in need of job placement assistance. A potential gap, therefore, is a lack of clear guidance on how to navigate this complex web of programs. Universal access to complete and updated information on federal programs and guidance for unique individual needs could be of great assistance to reserve component members.

Potential Gaps and Overlaps in Populations Served

In Table S.1 we also include information on whether the services provided by each program, resource, or office are targeted explicitly to the reserve components. Of all of the programs included, seven are targeted to the reserve components, and all of these targeted programs are offered by DoD. With the exception of the Yellow Ribbon Reintegration Program (YRRP)—which targets activated reserve component members—all of these reserve component-focused programs are designed to broadly serve all reserve component members. Programs in DoL generally serve all Americans, so they are available to all reserve component members, though American Job Centers provide some additional services to veterans. VA programs are often limited to members of the reserve components who are veterans, veterans with disabilities, or individuals utilizing education benefit programs.

In sum, our review of the existing portfolio suggests that the large number of federal employment programs and resources cover the full reserve component population (and many others as well). However, while many of these programs and resources are accessible to reserve component members, most of them are not targeted to or designed for the reserve components.

---

[4] U.S. Government Accountability Office, *Fragmentation, Overlap and Duplication: An Evaluation and Management Guide*, Washington, D.C., GAO-15-49SP, 2015; Ellen M. Pint, Amy Richardson, Bryan W. Hallmark, Scott Epstein, and Albert L. Benson, *Employer Partnership Program Analysis of Alternatives*, Santa Monica, Calif.: RAND Corporation, TR-1005-A, 2012.

Potential Areas of Concern

While we did not find any large gaps in the populations being served by federal employment programs, there are some populations that might have difficulty accessing the programs and services, and our informational discussions and prior research suggest that potential gaps might emerge if programs are consolidated. Four populations of reserve component members are of particular concern: (1) personnel without veteran status, (2) young members, (3) underemployed individuals, and (4) individuals who live far from military installations or in rural areas. If programs are streamlined or consolidated, special attention should be paid to these populations to ensure that they do not fall through the cracks.

Recommendations

Given our findings, we offer the following recommendations for improving federal job placement assistance for reserve component members:

- **Identify the employment needs of reserve component members.** We recommend that the federal government conduct a comprehensive assessment of the current employment needs of reserve component members. Without a clear articulation of the employment needs of reserve component members, it is impossible to determine whether program activities are effectively serving the needs of reserve component members. Future program evaluations or changes in programs should also be grounded in this assessment of the needs of reserve component members.
- **Evaluate the effectiveness of existing programs in meeting the needs of reserve component members.** Once the current employment needs of reserve component members are identified, we recommend that the federal government conduct in-depth program evaluations of the effectiveness of major programs across federal agencies in meeting the employment needs of reserve component members. Such formal program evaluations should include identifying program goals and outcomes (including the impact the programs have on increasing employment opportunities and results for reserve component members, as well as their effectiveness in meeting the reserve component needs identified through our first recommendation). These evaluations would also facilitate the identification of strengths and weaknesses within programs, as well as gaps and overlaps across federal programs. As with our first recommendation, conducting formal program evaluations should not be predicated on searching for cost reductions. Performing the evaluations themselves will cost money and take time. However, the important output of these program evaluations will be a comprehensive look at the current landscape of programs, mapped to the needs of the reserve components. Benefits will derive from avoiding duplicate, and possibly ineffective, programs in the future.
- **Assess the costs and benefits of streamlining existing programs and resources.** Based on the information collected in our first two recommendations, DoD and the rest of the federal government should then assess the costs and benefits of streamlining existing federal programs and resources. Both the needs assessment and the program evaluations will provide information that will help to determine whether streamlining programs and resources should be carried out, and, if so, which programs and resources should be streamlined. Streamlining could have many benefits. For instance, it could potentially

reduce costs by eliminating redundant services, tools, and resources. Streamlining could also help reserve component members identify the services they need more easily. However, before streamlining any overlapping services, it is critical to first identify whether those overlaps exist for a reason (e.g., to meet different needs or serve different populations). It is also important to recognize that while streamlining programs and resources could reduce salary costs, it could also limit service delivery to some populations. Therefore, it is essential to ensure that the needs of reserve component members continue to be met during any consolidation of programs.

- **Consider the pros and cons of moving primary responsibility for reserve component employment assistance from DoD to other federal departments.** Given that DoD's primary mission is warfighting, some might argue that it is appropriate to move primary responsibility for reserve component employment assistance away from DoD to other federal departments, such as DoL and the VA, whose primary missions more closely align with providing employment assistance to all Americans or social services to veterans. Simply adding reserve component employment needs to programs that already serve broader populations might save costs, streamline services, and potentially make it easier for service members to navigate employment resources.

  However, these potential benefits should be weighed carefully with the potential downsides of moving primary responsibility for reserve component employment assistance away from DoD to other federal departments. For instance, DoD has the best access to its reserve component population and can potentially collect vital data on the needs of this population easier than other federal agencies can. In addition, DoD has already built a sophisticated infrastructure (through its Hero2Hired program) to address the employment needs of reserve component members. The key issue is to identify whether there are unique reserve component needs that warrant a parallel set of employment services specifically for reserve component members.

  We recommend that if the primary responsibility for the administration of reserve component employment assistance is ultimately moved from DoD to another federal department, a DoD position should be established to monitor employment services for reserve component members in non-DoD federal departments. This should not simply be a liaison position, but rather a formal full-time position with the associated responsibilities, authorities, funding, and accountability to ensure that the employment needs of reserve component members are being met within the context of a broader employment program in a non-DoD agency. This could be a feasible option for delegating the administration of employment programs to non-DoD agencies, while maintaining some DoD oversight over their effectiveness in meeting the needs of reserve component members.

- **Make existing programs more applicable to the reserve components and increase awareness.** One of the most feasible, short-term steps that DoD (and the federal government as a whole) could take to improve employment assistance for reserve component members is to make the most of programs that already exist and increase awareness of those programs among reserve component members. This could prevent the proliferation of additional, potentially redundant programs. Most important, the federal government should consider making existing assistance (such as the Transition Assistance Program and American Job Centers) more applicable to reserve component members. This could increase program impacts, reduce costs, and provide consistent information across the components. Simply increasing awareness of these existing programs could potentially have a positive impact on reserve component members' employment outcomes.

- **Improve coordination and information-sharing across programs.** Finally, we recommend that the federal government improve coordination and information-sharing across employment programs. One of the major findings from our informational discussions is that some programs appear to have very little visibility on other programs' activities; thus, it is not a surprise that there are some overlaps across programs. Coordination could be increased quickly through very feasible, low-cost measures, such as periodic meetings among program staff, where they can share information on issues (e.g., the types of employment needs that they are seeing among reserve component members, what their programs are currently doing to address those needs, and any planned future changes in their programs' activities).

Things to Consider Before Making Changes

In light of our study findings, we also identify several items that DoD and the rest of the federal government should consider before making changes to the current federal portfolio of employment assistance to reserve component members. These include

- Assess the impact of potential changes to employment programs or activities.
- Ensure that new programs will not overlap with existing programs and activities.
- Recognize that cost is only one aspect of effectiveness.
- Plan up front to facilitate cost measurement and comparison across programs.

These measures are vital steps in facilitating the increased effectiveness of federal employment programs that assist reserve component members. These measures will also help to ensure that any changes to those programs' activities will be guided by the needs of reserve component members and that those changes will decrease duplication and instead fill potential gaps in services.

Acknowledgments

The authors would like to extend thanks to our sponsors at the Office of the Assistant Secretary of Defense for Reserve Affairs for their support for the research, especially Marie Balocki. Daniel Allen, Kevin Little, and Peter Weeks were also incredibly helpful in providing oversight of this research effort.

We also benefited from the contributions of RAND colleagues. Susan Gates and Laura Werber provided formal peer reviews that ensured that our work met RAND's quality standards. We also appreciated the helpful input from RAND management, including Jennifer Lamping Lewis, John Winkler, and Laura Baldwin. Michelle McMullen provided administrative support.

Finally, we note that we could not have completed this work without the participation of managers of federal employment programs in confidential discussions.

We thank them all, but we retain full responsibility for the objectivity, accuracy, and analytic integrity of the work presented here.

CHAPTER ONE

Introduction

Over the past few years, a proliferation of employment programs and resources has emerged for service members. However, job placement assistance and related employment services might be particularly critical to reserve component members because of some of the challenges that they face as they repeatedly transition between their military and civilian careers. For instance, while employer support for reserve component members is generally strong,[1] some employers might be concerned about hiring reserve component members because of their military obligations. This can be especially challenging for reserve component members following deployment.[2] For instance, a prior study found that some reserve component members struggled to find a new job after being demobilized, and some reserve component members indicated that they did not return to positions that they perceived to be equivalent to the ones held prior to deployment, that they did not receive a promotion, or that they lagged in some way with respect to pay raises or professional development.[3] Additionally, reserve component members might not return to the same employer after a military duty–related absence (either by choice or because of the employer's circumstances during deployment).[4]

Young reserve component members might be especially vulnerable to employment challenges. For instance, in his testimony to Congress in 2013, the director of Family and Employer Program and Policy, with the U.S. Department of Defense (DoD), indicated that while overall National Guard and Reserve unemployment rates were 13 percent in 2011 and 11 percent in 2012, during both years, the employment rate among junior enlisted ranks (e.g., E-1 to E-4) was significantly higher—23 percent in 2011 and 18 percent in 2012.[5] Research also indicates that some employment service providers find that young reserve component members might

---

[1]   Susan Gates, Geoffrey McGovern, Ivan Waggoner, John D. Winkler, Ashley Pierson, Lauren Andrews, and Peter Buryk, *Supporting Employers in the Reserve Operational Forces Era*, Santa Monica, Calif.: RAND Corporation, RR-152-OSD, 2013, p. 47.

[2]   Steve Vogel, "Returning Military Members Allege Job Discrimination—by Federal Government," *The Washington Post*, February 19, 2012; Tom Dreisbach and Rachel Martin, "National Guard Members' Next Battle: The Job Hunt," *Weekend Edition*, NPR.org, April 29, 2012.

[3]   Laura Werber, Agnes Gereben Schaefer, Karen Chan Osilla, Elizabeth Wilke, Anny Wong, Joshua Breslau, and Karin E. Kitchens, *Support for the 21st Century Reserve Force: Insights on Facilitating Successful Reintegration for Citizen Warriors and Their Families*, Santa Monica, Calif.: RAND Corporation, RR-206-OSD, 2013.

[4]   Gates et al., 2013, pp. 57–59.

[5]   Ronald Young, director, Family and Employer Program and Policy, U.S. Department of Defense, testimony to House Committee on Veterans' Affairs, March 14, 2013.

not have much experience presenting themselves to potential employers, and they might need more guidance in navigating the job search process.[6]

Another challenge that reserve component members face is that they are limited in their eligibility for employment support. For instance, many reserve component members are not eligible for some employment programs that are limited to those with veteran status.

Under federal law, the term *eligible veteran* means a person who

(A) served on active duty for a period of more than 180 days and was discharged or released therefrom with other than a dishonorable discharge; or

(B) was discharged or released from active duty because of a service-connected disability; or as a member of a reserve component under an order to active duty pursuant to section 12301(a), (d), or (g), 12302, or 12304 of title 10, served on active duty during a period of war or in a campaign or expedition for which a campaign badge is authorized and was discharged or released from such duty with other than a dishonorable discharge.[7]

Because reserve component members' time on active duty is, by its nature, limited, these requirements can be difficult to meet. As of March 2015, Employer Support of the Guard and Reserve (ESGR) estimated that there are about 700,000 reserve component personnel who did not qualify for veteran status. According to these estimates, there are more reserve component personnel who do not qualify as veterans than there are those that do qualify as veterans.[8]

Lastly, even when reserve component members are eligible for employment programs, they might have difficulty accessing them. In particular, some reserve component members do not live close to their units or to an active duty installation where some of these services are provided. This can cause challenges because they might not be aware of some services or cannot access some services.[9] Research has found that service providers try to address this issue by using targeted strategies to deliberately reach these geographically dispersed populations.[10]

In addition to these challenges faced by reserve component members, federal employment assistance to service members has also been undergoing a recent transition. In April 2014, First Lady Michelle Obama and Dr. Jill Biden announced the launch of the Veterans Employment Center (VEC) within the Department of Veterans Affairs (VA). The VEC is "a new integrated employment tool designed to connect veterans and service members with employers, and to help translate military skills into the civilian workforce."[11] The impetus behind this interagency effort was to improve, simplify, and consolidate the various employment resources for service members. In conjunction with the launch of the VEC, the online capabilities of DoD's primary employment program for reserve component members, the Hero2Hired (H2H) pro-

---

[6]   Werber, Wenger, et al., 2015.

[7]   United States Code (U.S.C.), Title 38, Section 4211, Definitions, January 3, 2012.

[8]   Discussion with federal program managers, March 31, 2015.

[9]   Paul D. Prince, "Out of Sight but Not Forgotten; Study Looks at Geographically Dispersed Soldiers, Families," Army. mil, October 7, 2009; Office of the Secretary of Defense for Reserve Affairs, *Benefits Guide*, Washington, D.C.: U.S. Department of Defense, September 2012.

[10]   Werber, Schaefer, et al., 2013.

[11]   Office of the First Lady, White House, "Obama Administration Launches Online Veterans Employment Center: One-Stop-Shop Connects Veterans, Transitioning Service Members, and Their Spouses to Employers," April 23, 2014.

gram, were transferred to the VA's VEC. This has raised a question among some about whether DoD should remain the primary proponent of job placement services and related employment assistance, or whether those services should be moved to non-DoD agencies.[12] It is within this transitional context that this study provides an assessment of the current array of employment assistance programs and resources available to reserve component members. This study also offers recommendations regarding (1) how to improve employment assistance to reserve component members and (2) what to consider before additional changes are made to program funding or activities.

The Office of the Secretary of Defense asked RAND to conduct the congressionally mandated study outlined in Section 583 of the 2015 National Defense Authorization Act (NDAA), which requires that "the Secretary of Defense . . . conduct a review of the feasibility of improving the efforts of the Department of Defense to provide job placement assistance and related employment services directly to members in the National Guard and Reserves."[13] The aims of this study were to identify the current universe of federal employment programs and resources available to reserve component members, assess the gaps and overlaps in those programs and resources, and develop recommendations for how DoD and the federal government as a whole can improve efforts to provide job placement assistance and related employment assistance to reserve component members. The study was carried out over the course of 11 weeks, from March 11, 2015, to May 27, 2015.

Study Approach

This study consisted of three tasks: (1) review the literature on federal job placement assistance and related employment services for reserve component members, (2) conduct informational discussions with managers of federal programs designed to provide job placement assistance and related employment services to reserve component members, and (3) assess the feasibility of improving DoD efforts to provide job placement assistance and related employment services to reserve component members. Since the study team could not find a compilation of all federal employment programs that provide employment services to reserve component members, and because our research sponsor was not aware of such a compilation, our study approach consisted of identifying the universe of employment programs and services available to reserve component members, identifying potential gaps and overlaps in those programs and services, and identifying potential ways to improve employment services to reserve component members.

Task 1 began with a comprehensive review of the literature on federal programs focused on providing employment-related services to members of the reserve components. The study team was able to leverage a large body of work that RAND has already conducted on improving reintegration support for reserve component members, as well as issues related to veterans' employment and unemployment. Appendix B provides the data-abstraction form that the research team developed to systematically collect information about federal programs that

---

[12] Chapter Four of this report discusses the pros and cons of transferring such services and resources to non-DoD agencies, versus retaining them within DoD.

[13] Public Law 113-291, Carl Levin and Howard P. "Buck" Mckeon National Defense Authorization Act for Fiscal Year 2015, Section 583, December 19, 2014.

provide employment assistance to reserve component members. Our review of the 40 programs identified in this study was based on 44 data sources. Those data sources can be found in Appendix D.

Task 2 entailed speaking with managers of federal programs that provide employment support to members of the reserve components. This allowed the research team to acquire supplemental information about federal employment programs that we could not acquire through the literature review conducted in task 1. We initially intended to conduct informational discussions with managers of all of the federal programs that provide employment assistance to reserve component members. However, as indicated in the caveats and limitations section below, we were limited to conducting discussions with nine federal managers. These discussions were held with managers of programs in DoD, the Department of Labor (DoL), and the VA that provide employment services to reserve component members. Our discussions were guided by the questions outlined in our discussion protocol (found in Appendix C of this report). The notes from these discussions were analyzed in two ways: (1) to identify program-specific insights and (2) to obtain general observations about the portfolio of employment support options.

Our findings from tasks 1 and 2 resulted in the development of a compendium of federal programs, resources, and offices that provide employment services to reserve component members. In Appendix A, we provide a summary table documenting some of the key characteristics of the programs, including populations served, services provided, and intensity of services. Our findings from tasks 1 and 2 also informed task 3 and our effort to identify gaps and overlaps in federal employment programs, resources, and offices that are available to reserve component members.

It is critical to note that a comprehensive gap analysis would require a comprehensive assessment of the employment needs of reserve component members, as well as evaluations of the programs that provide employment services to reserve component members. Since the study team could not find a comprehensive needs assessment, we instead identified employment needs that reserve component members have articulated in previous research, as well as common employment services provided to the general public to meet general employment needs. The resulting categories of services are

- assistance with résumé and interview preparation
- assistance with job search and placement
- assistance with job and career planning
- skills assessment and certification
- hosting job fairs
- assistance with legal issues
- assistance with entrepreneurship
- internships, apprenticeship, and training
- financial assistance
- access to information and tools.

We then accounted for whether a program, resource, or office offers each of these ten categories of services. In addition, during our information discussions, we asked stakeholders about their perspectives on potential gaps and overlaps in federal employment programs and services. Based on our review of programs and our discussions with stakeholders, we identified several areas of potential gaps and overlaps in federal employment programs.

Finally, our findings from all of the tasks informed our recommendations for how to improve federal efforts to provide job placement assistance to reserve component members and items to consider before making any changes to the current portfolio of federal employment programs available to reserve component members.

Caveats and Limitations

The timeline for this study shaped the scope and methodological approach in important ways. First, the findings are limited in their scope to the program data that were publicly available. As a means to double-check the accuracy of the information we found during our literature review, we originally planned to conduct informational discussions with the managers of all of the federal employment programs we identified. However, given DoD information-collection constraints under DoD Instruction 8910.01, *Information Collection and Reporting*,[14] we were not able to do so. If the research team had conducted more than nine informational discussions with DoD personnel, we would have had to acquire a report control symbol (RCS) license. However, given the very short timeline for this study, it was not feasible to acquire an RCS license; therefore, the research team conducted only nine informational discussions.

Additionally, as noted earlier, a complete analysis of gaps in services or an evaluation of the adequacy in employment assistance would require a comprehensive assessment of the employment needs of reserve component members. Given the absence of such an assessment, we instead identified potential gaps in services by assessing federal programs against employment needs that have been articulated by reserve component members in prior research, as well as common employment-related services for the general public. As a result, our findings regarding potential gaps in employment services are quite limited, and they do not take into account the full range of employment needs among the reserve components.

In addition, our current analysis describes the supply of programs, but it cannot compare these programs with demand for employment services or their adequacy in addressing the needs of reserve component members. While we identify areas of overlap and areas where fewer services are provided and point decisionmakers to specific areas of concern, our analysis is not sufficient to determine whether existing programs are meeting reserve component members' needs and which programs are the most effective and efficient at doing so. To determine which programs are best meeting the needs of reserve component members, full program evaluations would be necessary.

It is also important to note that our analysis of gaps and overlaps focuses only on federal programs. State programs and nongovernmental organizations also offer a range of employment services for reserve component members that might help to fill gaps and add additional areas of overlap.

Lastly, while we were able to identify most programs' eligible populations, program services, and the intensity of those services, we were not able to evaluate the effectiveness of those activities or the programs in general. As indicated in our recommendations in Chapter Four, a

---

[14] U.S. Department of Defense Instruction 8910.01, Information Collection and Reporting, Washington, D.C.: U.S. Department of Defense, May 19, 2014.

formal evaluation of the effectiveness of federal employment programs would be a critical step in identifying how DoD and the federal government as a whole could improve job placement assistance and related employment services to reserve component members.

Therefore, this study represents a preliminary assessment of this important topic, and there are opportunities to expand and enhance the analysis in the future. For instance, with a longer timeline, it would be feasible to acquire an RCS license and conduct more interviews with directors of federal programs that employ National Guard and Reserve members. In addition, a comprehensive needs assessment would enable future analyses to a conduct a complete assessment of gaps in services or an evaluation of the adequacy in employment assistance for reserve component members.

Study Scope

There are numerous other employment programs and resources that we did not include because we considered them outside the scope of this study. For example, because the NDAA language specifically focused on assistance to reserve component service members, we did not include programs or resources that exclusively provide services to reserve component spouses or active duty service members but not reserve component service members. We also excluded state or nongovernmental programs and resources because the language in the NDAA specifically required our study to focus on federal employment assistance. Finally, we excluded programs, resources, and offices that are primarily aimed at providing services and resources to employers. These include such resources as the VA Veterans Employment Toolkit.

Organization of This Report

Chapter Two presents an overview of federal employment programs, resources, and offices that are available to reserve component members. The chapter provides brief descriptions of the federal programs, resources, and offices that we identified as providing job placement assistance and related employment services to reserve component members. We have grouped the programs, resources, and offices according to the federal departments or agencies responsible for overseeing them.

Chapter Three presents an overview of our analysis of the gaps and overlaps in federal programs, resources, and offices available to reserve component members. For each program and resource, we identified the types of services provided, the intensity of those services, and the populations that are served. Key stakeholders from each of the major agencies that provide employment programs to reserve component members also shared with us basic information about the programs, provided feedback on our full list of programs to identify gaps in our data, and shared some perspectives regarding where gaps and overlaps might exist and how they might be addressed. Based on our review of programs and our discussions with stakeholders, we identify several areas of potential gaps and overlaps in federal employment programs.

Chapter Four presents our recommendations and their implications for improving DoD and federal job placement assistance, as well as related employment services and resources, for reserve component members. In addition, this chapter identifies items to consider before making additional changes to program funding or activities. Lastly, this chapter lays out a road map for improving federal job placement assistance, and it discusses the pros and cons of transferring such services and resources to non-DoD agencies.

In Appendix A, we provide a summary table documenting some of the key characteristics of the programs, including population served, services provided, and intensity of services. Appendix B presents the data-abstraction form that we developed to systematically collect information about federal employment programs, resources, and offices that provide employment assistance to reserve component members. Appendix C presents the protocol that we developed to guide our discussions with stakeholders. Appendix D presents our data sources for the 40 programs in our study.

CHAPTER TWO

Overview of Federal Employment Programs for Reserve Component Members

In this chapter, we provide summary descriptions for the federal programs, resources, and offices that we identified as providing job placement assistance and related employment services to reserve component members. We describe the approach we took to document these programs, resources, and offices and then present them by their overseeing agencies, including DoD, DoL, the VA, and several other federal agencies. In Appendix A, we provide a summary table documenting some of the key characteristics of the programs.

Approach to Documenting Programs, Resources, and Offices

To document the federal programs, resources, and offices available to reserve component members for job placement assistance and related employment services, we used a systematic and varied set of approaches. We started by conducting a broad literature review and using general Internet search engines (including Google), as well as the search engines embedded in federal agency and military service websites, to search for the terms *reserve component* and *employment*. The program websites also often have menus of or links to other employment programs that we were able to mine to identify new programs. In addition, through our conversations with a group of key stakeholders in federal agencies, we received some recommendations for programs to include and some additional information about employment programs for reserve component members. Finally, we reviewed recent literature to identify additional information about programs that serve the reserve components. These searches were guided by a data-abstraction form that allowed us to document information systematically across programs. This data-abstraction form is included in Appendix B.

For the purposes of this study, we define a *program* as a set of activities that provides direct services, tied together through shared assets (e.g., staff, funding, space, materials), meant to affect a targeted population's knowledge, attitudes, or behavior to accomplish a specific goal (or goals).[1] Such programs typically include some aspect of human interaction. We also found resources that might not technically be defined as programs but are important components of the federal portfolio. These websites and tools are regularly used by service members as part of their job searches and likely require federal funds to maintain. In addition, we also include

---

[1] This definition is loosely based on what was developed in RAND's Program Classification Tool. See Joie D. Acosta, Gabriella C. Gonzalez, Emily M. Gillen, Jeffrey Garnett, Carrie M. Farmer, and Robin M. Weinick, *The Development and Application of the RAND Program Classification Tool: The RAND Toolkit, Volume 1*, Santa Monica, Calif.: RAND Corporation, RR-487/1-OSD, 2014.

several offices that might not themselves be considered a program but that typically oversee programs and also provide additional services and resources to service members.

As indicated in Chapter One, there are many other programs, resources, and offices that provide job placement assistance and related employment services to reserve component members that we did not include because we considered them outside the scope of the study. For example, because the NDAA language specifically focuses on assistance to reserve component service members, we did not include programs or resources that exclusively provide services to reserve component families and spouses or active duty service members. We also excluded state or nongovernmental programs, resources, and offices because the language in the NDAA specifically required our study to focus on federal employment assistance. Finally, we excluded programs, resources, and offices that are primarily aimed at providing services and resources to employers. Table 2.1 lists the programs that we identified, by agency, that provide employment assistance to reserve component members.

Table 2.1
Federal Job Placement Assistance Available to Reserve Component Members

Agency	Program Name
DoD	Airman and Family Readiness (Air Force)
	Always a Soldier (AAS) (Army)
	Beyond Yellow Ribbon (BYR) programs
	Career Skills Program (CSP) (Army)
	Credentialing Opportunities Online (COOL) (Air Force/Army/Marine Corps/Navy)
	DoD Hiring Heroes
	Education and Employment Initiative
	Employment Readiness Program (ERP) (Army)
	Fleet and Family Support Centers (Navy)
	Guard Apprenticeship Program Initiative (GAPI) (Army National Guard)
	Hero2Hired (H2H)
	Job Connection Education Program (JCEP) (National Guard)
	Marine and Family Program's Career Services (Marine Corps)
	Military OneSource
	National Guard Employment Network (NGEN) (National Guard)
	Operation Warfighter (OWF)
	Partnership for Youth Success (PaYS) (Army)
	Public Private Partnership (P3) Program (Army Reserve)
	SkillBridge
	Troops to Teachers (TTT)
	Yellow Ribbon Reintegration Program (YRRP)

Table 2.1—Continued

Agency	Program Name
DoL	American Job Centers
	Career OneStop
	O*NET and My Next Move for Veterans
	Registered Apprenticeship
	Veterans' Employment and Training Service (VETS)
VA	Compensated Work Therapy (CWT)
	Education and Career Counseling and CareerScope
	Office of Small and Disadvantaged Business Utilization (OSDBU)
	VA for Vets
	Veterans Employment Center (VEC)
	Veterans Opportunity to Work (VOW)
	VetSuccess on Campus (VSOC)
	Vocational Rehabilitation and Employment (VR&E)
	Workstudy Program
Office of Personnel Management (OPM)	Feds Hire Vets
DoD, DOL, VA	National Resource Directory
	Transition Assistance Program (TAP)
Small Business Administration (SBA)	Veterans Business Outreach Program (VBOP)
All federal agencies	Veterans Employment Program Offices (VEPOs)

The remainder of this chapter provides a summary of each program in Table 2.1, including the program's purpose, which agency has oversight over it, the program's eligible and target populations, and the services that the program provides.

Department of Defense

DoD, the National Guard, and the services offer a range of programs, resources, and offices to reserve component members to assist with job placement and related employment services. A full list of programs across all agencies is presented in Table 2.1. The programs, resources, and offices overseen by DoD are typically targeted to current or transitioning service members as opposed to veterans, and sometimes targeted to veterans and the families of service members. Several of the programs are explicitly designed for and targeted to the reserve components. In this section, we first document the programs, resources, and offices that are offered by the Office of the Secretary of Defense, and then we discuss those offered by the National Guard, the Reserves, and the services.

Airman and Family Readiness (Air Force)

Purpose: Airman and Family Readiness serves as a one-stop information and referral center to ensure that personnel and their families are connected with the appropriate services, on and off base.

Oversight: The program is administered by the Air Force.

Population: Airman and Family Readiness is targeted to active duty service members, reserve component members, civilians, retirees, and their families.

Services: The program's website describes numerous different services and resources provided by the program, including financial assistance, crisis assistance, and employment support. In the area of employment support, the program provides résumé and interview assistance and support with job searches and networking. The website information suggests that one-on-one services might be available, but there is no information about contacts or methods of connecting to resources. In addition to a list of available services and resources provided by the program, the website offers links to other programs and tools (e.g., Military OneSource, and YRRP). The site provides a link to the *Personal and Family Readiness Handbook*, but the link to the handbook was broken when we tried to access it on July 30, 2015.[2]

Always a Soldier Program (Army)

Purpose: The goal of the program is to provide continuing support to individuals beyond their active duty service and to assist wounded veterans after their service to the United States and its allies.

Oversight: Always a Soldier (AAS) is overseen by U.S. Army Materiel Command.

Population: Veterans with a 30-percent or greater service-related disability who are eligible for the Veterans Recruitment Appointment or the noncompetitive hiring authorities for 30-percent disabled veterans.

Services: The exact services provided are tailored to individual participants. According to program guidance documentation, the services cannot be tailored by major subordinate commands. However, none of the resources we were able to find describes the typical services provided by the program, outside of providing individuals with priority consideration for jobs in the Army Materiel Command. Potential employment tracks include wage grade trades and labor positions, internships, and full General Schedule positions. While finding jobs is the primary service, one resource noted that the program also assists veterans and their family members in purchasing food and clothes, explaining job application processes, and providing resource information for veterans.[3]

Beyond Yellow Ribbon Programs

Purpose: Beyond Yellow Ribbon (BYR) programs are federally funded state programs designed to supplement the services that are provided by the Yellow Ribbon Reintegration Program (see the description below) and to extend support to reserve component members and their families after the deployment cycle.

---

[2]   For more information on the Air Force's Airman and Family Readiness program, see "Air Force Airman and Family Readiness," web page, Air Force Reserve Command, n.d.

[3]   For more information on the AAS program, see "Army Always a Soldier," web page, U.S. Army Materiel Command, n.d.; the resource describing other services was found at "Army Materiel Command's Always a Soldier Program," web page, Career Center for Wounded Warriors and Disabled Veterans, Military.com, n.d.

Oversight: The state programs are administered through a mix of nonprofits, the National Guard, and state agencies. The federal program is overseen by the Office of the Secretary of Defense.

Population: The programs are targeted to reserve component members and their families.

Services: Federal funding is provided to state-level programs that offer services in such areas as employment, mental health, and legal assistance. For the most part, services provided by BYR programs are one-on-one high-touch services that involve case management. Many BYR employment-focused programs provide assistance with résumés and interviewing, access to job postings, and communication with employers.[4]

Career Skills Program (Army)

Purpose: The Career Skills Program provides such opportunities as apprenticeships, on-the-job training, job shadowing, employment skills training, and internships for service members preparing to transition from military to civilian employment.

Oversight: The program is overseen by U.S. Army Human Resources Command.

Population: Service members must have a minimum of one year of active duty service or two years of reserve component service at the time when the course starts or the credential or licensing exam is administered. Individuals must also have completed at least 180 continuous calendar days of active duty service and must expect to be discharged or released from active duty within 180 calendar days of the commencement date of participation in the program. While the program is targeted to the Army, assistance is also available to individuals in the other services.[5]

Services: The Career Skills Program provides funding to cover the costs of credentialing opportunities as long as the opportunity meets several requirements, including that the credentials are tied to the service member's previous duties and should be relevant or applicable to current or future needs of the unit or the Army, the credential is recognized by the Army training school or center, and the individual meets the requirements of the credentialing organization.

Credentialing Opportunities Online (Air Force, Army, Marine Corps, and Navy)

Purpose: Credentialing Opportunities On-Line (COOL) provides information and funding support for individuals who are interested in earning certifications and licenses.

Oversight: While COOL was initially developed by the Army and targeted to individuals in the Army, COOL programs are also offered by the Air Force, Marine Corps, and Navy.

Population: The website advertises the tool as being useful for service members; counselors assisting with education, careers, and transition assistance; Army recruiters; employers; and credentialing boards.

Services: COOL provides a search engine that helps service members find information about certifications and licenses related to their Military Occupational Specialties (MOSs). The tool includes information about eligibility requirements, exam topics, contact information, and links to the credentialing agency's website. Information also includes whether the credential is in a rapidly growing occupation, whether the costs can be covered by the GI Bill or Army e-Learning Center, whether the credentialing opportunity counts toward promotion

---

[4]   For more detail on BYR programs, see Werber, Wenger, et al., 2015.

[5]   For more information on the Army's Career Skills Program, see "Career Skills Program—CSP," web page, United States Army Human Resources Command, n.d.

within the military, instructions about how to get funding, and whether the credentialing opportunity is accredited by various agencies (e.g., the American National Standards Institute). In addition to informational assistance, some of the services' programs offer funding to cover credentialing costs.[6]

DoD Hiring Heroes

Purpose: DoD Hiring Heroes provides individuals with resources and services that connect them with employment in DoD, other federal agencies, and private-sector corporations.

Oversight: The program is administered by the Defense Civilian Personnel Advisory Service.

Population: DoD Hiring Heroes serves wounded, ill, injured, and transitioning service members, veterans, spouses, and primary caregivers. The program is not explicitly targeted to the reserve components.

Services: The program focuses on organizing and conducting specialized career fairs to increase awareness of job opportunities. The Recruitment Assistance Division (RAD) offers networking events specifically designed to help severely injured service members transitioning to careers in government and the private sector. Services offered at the networking events include career guidance, résumé-writing workshops, and interviewing-skills training. Individuals are provided with a toll-free number that allows them to speak directly to RAD career advisors for guidance on job search processes, information about current vacancies, and assistance in completing applications. In addition to accessing advisors by phone, service members and other eligible populations are provided with weekly opportunities to connect with advisors in one-on-one online chat sessions, and individuals can connect with advisors by email. The website provides a calendar of upcoming career events, access to such resources as a military-to-civilian skill translator and information on veterans' preferences, and links to other programs that provide employment services.[7]

Education and Employment Initiative

Purpose: The Education and Employment Initiative provides individuals assistance with identifying employment skills and provides education and career planning.

Oversight: The program is overseen by the Office of Warrior Care Policy.

Population: The Education and Employment Initiative serves active duty and reserve component service members recovering from service-related injuries. It is not explicitly targeted to the reserve components.

Services: The program has ten regional coordinators who are tasked with working across the nation with military departments, federal agencies, the private sector, and institutions to locate training, employment, and education opportunities for recovering service members. The coordinators work individually with service members to develop career decisions; plans for after secondary, graduate, or professional school; employment plans; or job search competencies.[8]

---

[6]   For more information on COOL, see Credentialing Opportunities On-Line, website, n.d.

[7]   For more details on DoD's Hiring Heroes program, see "DoD Hiring Heroes Program," web page, U.S. Department of Defense, n.d.

[8]   For more information on DoD's Education and Employment Initiative, see "Education and Employment Initiative," web page, Warrior Care Blog, DoD Office of Warrior Care Policy program website, n.d.

Employment Readiness Program (Army)

Purpose: The purpose of the Employment Readiness Program (ERP) is to provide job search assistance and referral services.

Oversight: ERP is offered through the Army Community Service (ACS) centers and over-seen by the Army.

Population: The program is available for service members, including active duty service members, reserve component members, retirees, and DoD civilians, as well as military spouses. The program is not explicitly targeted to the reserve components.

Services: The program offers one-on-one job search assistance by a professional job search trainer at an ACS center. The website provides a list of ERP managers for individuals to contact. ERP provides access to online databases that list employers and employment opportunities, as well as employment information from state, local, and government agencies. ERP also provides classes and seminars on a range of different topics, including self-assessment and career exploration, résumé writing, interviewing techniques, dressing for success, networking, and entrepreneurship. Job search trainers also offer résumé critiques, career counseling, and computers for use in job searches. ERP also provides access to job fairs and hiring events.[9]

Fleet and Family Support Centers (Navy)

Purpose: Fleet and Family Support Centers (FFSCs) provide a full range of quality products and services in support of mission readiness and retention and to oversee policy development, resourcing, and the oversight of quality-of-life programs.

Oversight: The centers are overseen by the U.S. Navy.

Population: The program serves members of the Navy, including active duty members, reservists, military retirees, and their families. Services are not explicitly targeted to reservists.

Services: FFSCs are located on Navy installations, and the centers offer a range of different courses on career readiness, including courses on entrepreneurship, résumé writing, and career training. In addition to the courses and workshops, counselors at the centers provide one-on-one employment assistance to service members and their families. FFSCs are also responsible for Transition GPS planning[10] through the Transition Assistance Program (TAP; described in detail later in the chapter). As part of Transition GPS, the Navy offers workshops through the FFSCs to provide information and guidance on career planning and access to veterans' benefits. Transition GPS workshops are only available to individuals by order after a deployment or mobilization. Other services offered by the FFSCs include counseling for personal or financial issues, relocation assistance, sexual assault services, and services for retired service members. The centers also provide access to computers, the Internet, and job postings for use in job searches.[11]

---

[9] For more information on the Army's ERP, see "Employment Readiness Program," web page, Army OneSource, n.d.

[10] Transition GPS (Goals, Plans, Success) is TAP's outcome-based, modular curriculum with standardized learning objectives.

[11] For more information on the Navy's FFSCs, see "Fleet and Family Support Centers," web page, Commander, Navy Installations Command, n.d.

Guard Apprenticeship Program Initiative (Army National Guard)

Purpose: The Guard Apprenticeship Program Initiative (GAPI) provides opportunities to attain national occupational skill certification through civilian apprenticeships. The goal of the program is to develop a more mission-capable force by educating military personnel.

Oversight: While the program is overseen by the National Guard, it was developed in collaboration with DoL and the VA.

Population: The program is targeted to individuals in the Army National Guard and Army Reserve.

Services: GAPI combines classroom studies with on-the-job training overseen by a trade professional or supervisor. The apprenticeships take one to five years to complete, and completion results in nationally recognized certification of apprenticeship in a specific skill. The occupations for which apprenticeships are offered include mechanics, medical technicians, therapists, truck drivers, construction engineers, website developers, law enforcement and military police officers, and culinary artists and cooks, among others.[12]

Hero2Hired

Purpose: H2H was created to address unemployment and underemployment within the reserve components so that the nation can maintain capabilities that respond effectively to national emergencies and wartime requirements.

Oversight: The program currently operates within the Family and Employer Programs and Policy within DoD, under the Employment Initiative Program, which is part of the Office for Reintegration Programs.

Population: H2H serves all reserve component members.

Services: The program provides high-touch, one-on-one support to individuals through employment coordinators (ECs), who are located in each state. ECs manage cases of individuals and provide assistance with many aspects of an individual's job search, including review and refinement of résumés, posting of open positions, preparation for interviews, and follow-up with employers to gather feedback about the application process. ECs also work to increase and maintain contact with potential employers that might want to hire members of the reserve components, and ECs use software to track job opportunities by region and keep a database of job seekers and their characteristics.[13] The H2H website includes contact details for H2H ECs and links to other employment-related services and resources for reserve component members. Until recently, H2H was responsible for maintaining a portal for job seekers and employers, but these capabilities were transitioned to the Veterans Employment Center (operated by the VA). H2H is supported by the Employer Support of the Guard and Reserve program (see the description below).[14]

---

[12] For more information on the National Guard's GAPI program, see "Guard Apprenticeship Program Initiative (GAPI)," web page, July 8, 2011.

[13] The efforts of the Army Reserve's Employer Partnership Office, ESGR, and Yellow Ribbon Reintegration Program for online job search capability and case managing were combined using H2H in March of 2013. The H2H website was later transferred to the VEC in September 2014, and the case management capabilities now being used by the VEC are those originally developed for the Wounded Warrior Project. For more information on DoD's H2H program, see Hero2Hired, website, n.d.

[14] The ESGR program was designed to develop and promote supportive work environments for members of the reserve components. While some do consider this an employment program for the reserve component (see U.S. Government Accountability Office, *Fragmentation, Overlap, and Duplication: An Evaluation and Management Guide*, Washington, D.C.,

Job Connection Education Program (National Guard)

Purpose: The Job Connection Education Program (JCEP) was developed by the National Guard to assist individuals who are unemployed or underemployed in finding careers.

Oversight: The program is overseen by the Army National Guard. The program initially began in Texas and has now expanded to Iowa, Wisconsin, and Tennessee.

Population: JCEP is targeted at service members from the National Guard and Reserves and their spouses.

Services: The program offers one-on-one counseling, education and training assistance, résumé and interview guidance, an online job search engine, and direct activities to connect employers in the region to JCEP and its participants. Services are provided to service members and their families through training and development counselors, and a business advisor takes the lead role in coordinating with employers. Individuals can register for JCEP services directly on the program website by entering information about location, military experience, job interests, and contact information. The registration form requires individuals to certify that they are service members who are currently serving or have separated within the past two years, or that they are the spouses of current service members. Individuals who register for JCEP services will then be contacted by counselors for additional information and one-on-one services.[15]

Marine and Family Program's Career Services (Marine Corps)

Purpose: The Marine and Family Program's Career Services program offers comprehensive information and resources concerning employment, training, education, and special employer events.

Oversight: The program is administered by the U.S. Marine Corps.

Population: The program is intended to serve all service members, retirees, veterans, DoD civilian employees, and their families. The program is not explicitly targeted to the reserve components.

Services: Marine and Family Program's Career Services offices are housed on Marine Corps installations. The program's Career Centers offer comprehensive information and resources concerning employment, training, and education. The Career Centers are staffed by career specialists, individuals who provide one-on-one services, including assistance in designing a résumé and building a LinkedIn account, developing interviewing skills, and searching for employment. In addition to career specialists, the Career Centers provide access to computers and printers, resource libraries, and special employer events. Centers on each installation offer their own websites, and these websites describe the center's resources, provide contact information for career specialists, and offer a calendar of upcoming events (e.g., workshops and employer events). The websites also provide links to other employment-related programs and resources, including the VA's eBenefits website, Office of Personnel Management's (OPM's) Feds Hire Vets program, and DoL's Career OneStop program.[16]

GAO-15-49SP, 2015), we do not include it as a program under this study. Our decision to exclude the program is due to the fact that it primarily targets services to employers as opposed to service members. For more information on ESGR, see Employer Support of the Guard and Reserve, website, n.d.

[15] For more information on the National Guard's JCEP program, see National Guard Job Connection Education Program, website, n.d.

[16] For more information on the Marine and Family Program's Career Services, see "Career Services," web page, Marine and Family Programs, Marine Corps Community Services Camp Pendleton, n.d.

Military OneSource

Purpose: Military OneSource provides an extension of services that are typically offered to service members who are not directly located on or near a military installation.

Oversight: The office of Military Community and Family Policy oversees the program.

Population: Military OneSource is available to service members in active duty, as well as the reserve components, regardless of activation status. The program is not explicitly targeted to the reserve components. The program also provides information for employers, military leadership, and service providers.

Services: The program provides a call center and Internet-based support, personal non-medical counseling, help with income taxes and other financial services, spouse education and career support, educational materials, and a social media hub. Individuals can connect to consultants by phone, in person, or over the Internet. The website provides a toll-free number for individuals to connect directly with a Military OneSource consultant, as well as contact numbers for specialists in the areas of crisis and sexual assault. To access services online, users are required to register and log in.[17]

National Guard Employment Network (National Guard)

Purpose: The National Guard Employment Network (NGEN) increases job and career match opportunities and promotes readiness, resilience, and retention among service members and their families.

Oversight: NGEN is administered by the National Guard. In some cases, the website appears to be managed directly by NGEN, while in other cases, the website directs users to other state-provided programs (e.g., Work for Warriors in California).

Population: NGEN resources are specifically intended for guard members and their families, though some of the NGEN-affiliated programs also provide services to reservists and veterans of active duty service.[18]

Services: After registering on the NGEN website, job seekers are provided with one-on-one assistance by NGEN counselors or state employment office staff members. The services provided to job seekers include assistance with résumé development and interview preparation, certification and licensing opportunities, military-to-civilian skill crosswalks, and access to job listings. The program also provides employers with the opportunity to register and post available job listings in the NGEN system. The website describes a process of "quality control" in matching service members to open positions, and the program provides case management to employers and job seekers. Service providers can also utilize the NGEN resources to support service members and their families through other programs. The resources offered to service providers include the NGEN database for case management and tracking and access to counselors and training. The website directs individuals to state-specific websites.

Operation Warfighter

Purpose: Operation Warfighter provides service members valuable work experience during their recovery and rehabilitation. The goal is to assist with the reintegration to duty or the transition

---

[17] For more information on Military OneSource, see Military OneSource, website, n.d.

[18] For more information on NGEN, see "National Guard Employment Network," web page, Nation Guard, n.d.

into the civilian work environment, where individuals are able to employ their newly acquired skills in a nonmilitary work setting.

Oversight: The Operation Warfighter program is overseen by DoD's Office of Warrior Care Policy.

Population: The program is targeted to wounded, ill, and injured service members in active duty and the reserve components. While the program is available to members of the reserve components, it is not explicitly targeted to them.

Services: Operation Warfighter is staffed by ten regional coordinators across the country, who work with eligible participants to explore career interests, build résumés, and provide opportunities for additional training, experience, and networking. Based on skills, expertise, and interest, individuals selected as "warfighters" are placed in host offices and assigned a supervisor and a mentor. Host offices are located at several federal agencies and include positions at the Department of Homeland Security, the U.S. Army Corps of Engineers, and DoD. Mentors help to familiarize participants with their assigned offices and the departments, provide guidance on how to perform specific work duties, and answer any questions that come up. Schedules are designed to accommodate medical treatments; individuals work approximately 20 hours per week for three to five months and receive free transportation to work. While there are no guarantees of employment after completion, individuals are considered potential candidates for full-time positions.[19]

Partnership for Youth Success (Army)

Purpose: The Partnership for Youth Success (PaYS) program is designed to support the recruitment of new enlistees in the Army.

Oversight: In April 2011, the U.S. Army Marketing and Research Group, a field-operating agency of the Office of the Assistant Secretary of the Army for Manpower and Reserve Affairs, assumed the responsibility for administering and managing the PaYS program.

Population: The program was initially developed for active duty Army members but was expanded to the Army National Guard and Army Reserve in 2002. Active duty service members are eligible after their first terms of service, while reserve component members are eligible immediately after completing basic training and job training.

Services: The program provides enlistees with the promise of a job interview and possibly a job with a civilian employer at the time of separation from the military. The program partners with a range of employers, including private industry, academia, and state and local government agencies. These partner employers commit to providing new service members with an interview opportunity, and to qualify for the program, employers must have at least 500 employees and available positions for full-time employment. According to the PaYS website, positions are loaded into the program database on an annual basis each July. In addition to providing general information about the program, the website provides testimonials from beneficiaries of the PaYS program. A help desk with a call-in line is available as a means for accessing additional information and setting up an account to log in to the program database.[20]

---

[19] For more information on Operation Warfighter, see "Operation Warfighter," web page, *Warrior Care Blog*, DoD Office of Warrior Care Policy, n.d.

[20] For more information on the Army's PaYS program, see Army Partnership for Youth Success, website, n.d.

Private Public Partnership (Army Reserve)

Purpose: The goal of the Private Public Partnership (P3) program is to establish and utilize relationships between public and private organizations to find employment for members of the Army reserve components and provide private organizations with an opportunity to benefit from the skills and experiences people gain in the military.

Oversight: Created under the Army Reserve in 2008 as the Employer Partnership Office, in 2013 the Private Public Partnership Office (PPPO) was launched to oversee the program.

Population: P3 is targeted to all reserve component members (veteran and nonveteran), as well as their families.

Services: The program provides training opportunities and the ability for reserve component members to apply their expertise and leadership skills to real-world projects. The PPPO focuses on three lines of services: individual readiness, leader readiness, and unit readiness. Since the PPPO began, it has grown from 27 sites to 65. Army career employment specialists (ACESs) are located in Army Reserve communities and work with soldiers and employers in local areas. Transition employment liaisons are based in transition centers on active duty bases and provide a handoff to an ACES. ACESs brief Army reserve units on employment opportunities, help with registering for job portals, and provide assistance in the applications process. Individuals seeking employment can also access subject-matter experts in the Fort Belvoir, Virginia, main program office, in one of eight areas: signal information technology, logistics, human resources and finance, medicine, law enforcement, education, public affairs, and engineering. Employment assistance includes résumé and interview preparation, access to information about employment opportunities, and assistance connecting to jobs and completing the application process.[21]

SkillBridge

Purpose: SkillBridge promotes civilian job training available for transitioning service members, including apprenticeships and internships.[22]

Oversight: The program is overseen by the Office of the Under Secretary for Personnel and Readiness, Assistant Secretary of Defense for Force Readiness and Training.

Population: Both training providers and service members must be approved for Skill-Bridge participation by the commanders at military installations and SkillBridge. The program is available to veterans without a dishonorable discharge and is not explicitly targeted to the reserve components.

Services: The program offers a range of apprenticeship and internship opportunities. The training can take place up to six months prior to a service member's separation. To qualify, the training must offer a high probability of employment and be provided to the service member at little or no cost. Service members use the SkillBridge application to search for training opportunities that best fit their goals, based on skill set, desired location, and transition date. When

---

[21] For more information on the Army Reserve's P3 program, see "Private Public Partnerships," web page, U.S. Army Reserve, n.d.

[22] For example, SkillBridge's website mentions that nearly 750 service members have participated in the United Association of Journeymen and Apprentices of the Plumbing and Pipefitting Industry's Veterans in Piping training program for welding and pipefitting skills. Microsoft has also participated in the program and provides information-technology training through Microsoft Academies at several installations.

service members find relevant training opportunities, they can inquire with the training provider to learn more details about the opportunity and the application process.[23]

Troops to Teachers

Purpose: To help eligible military personnel begin new careers as teachers in public schools where their skills, knowledge and experience are most needed.

Oversight: Troops to Teachers (TTT) was established in 1994 as a DoD program, oversight was transferred to the Department of Education in 2000, and then the program was transferred back to DoD in 2013. TTT is now managed by the Defense Activity for Non-Traditional Education Support (DANTES), in Pensacola, Florida.

Population: All current and former service members with honorable service are able to participate in the program. The reserve components are not explicitly targeted by the program, but reserve component members are eligible.

Services: The program has built a network of state TTT offices that provide participants with counseling and assistance regarding certification requirements, routes to state certification, and employment leads. The TTT homepage provides information and resource links, including links to state departments of education, state certification offices, and other job listing sites in public education. For some participants, financial assistance might be provided for the transition to the classroom. The financial assistance is for those becoming first-time teachers and is not available to those who became teachers prior to registration with TTT.[24]

Yellow Ribbon Reintegration Program

Purpose: The purpose of the Yellow Ribbon Reintegration Program (YRRP) is to promote the well-being of reserve component members, their families, and their communities by connecting them with resources throughout the deployment cycle.

Oversight: The Office of the Secretary of Defense's Office of Reintegration Programs oversees the program. Commanders and leaders also play a critical role in ensuring that reserve component members and their families attend YRRP events.

Population: The program is specifically targeted to reserve component members and their families, because, as the program's website emphasizes, "reintegration during post-deployment is a critical time for members of these populations, as they often live far from military installations and other members of their units."

Services: Through YRRP events, reserve component members and their families connect with local resources before, during, and after deployments. Resources provided at YRRP events and on the program's website include information on health care, education and training opportunities, and financial and legal benefits. Courses are taught at YRRP events, and these courses touch on such topics as mental health, parenting, and employment-related issues. Employers and DoD employment service providers from programs like ESGR and H2H often attend YRRP events to connect transitioning service members to employment resources.[25]

---

[23] For more information on SkillBridge, see DoD SkillBridge, website, n.d.

[24] For information on Troops to Teachers, see Troops to Teachers, website, n.d.

[25] For more information on DoD's YRRP, see "Yellow Ribbon Reintegration Program," web page, Joint Services Support, n.d.

Department of Labor

DoL also offers a range of job placement and related employment programs and resources that are available to reserve component members. The FY 2016 budget includes $500 million to expand all DoL employment programs.[26] While reserve component members are eligible for some DoL employment programs, none of the programs explicitly targets services exclusively to reserve component members. For the most part, the programs and resources are offered to all Americans, including civilians and nonveterans. Some of the programs target services to veterans, and reserve component members are eligible for these only if they qualify as veterans. We describe each of these programs and resources below.

American Job Centers

Purpose: American Job Centers (AJCs) provide a single access point to federal programs and local resources to help individuals find jobs, identify training programs, and gain skills in growing industries.

Oversight: The program is overseen by DoL.

Population: AJCs provide services to both job seekers and employers, which are available to all Americans (regardless of whether they are a veteran). Unemployed post–9/11-era veterans are given special access to services through the Gold Card program, which is overseen by the Veterans' Education and Training Service (VETS) agency. Nonveteran reserve component members are eligible to use some AJC services but are not explicitly targeted.

Services: Core AJC services include determination of eligibility for services; initial assessment of skills, abilities, aptitudes, and service needs; job search and placement; workforce information; and job placement follow-up. AJCs' employer services include workforce information, job description writing, posting job openings, reviewing applicants' résumés, organizing job fairs, skill upgrading and career ladders, offering places to conduct interviews, prescreening job applicants, assessments of applicants' skills, and referrals of job-ready candidates. AJCs also offer training in occupational skills, skill upgrading and retraining, on-the-job training, workplace training and related instruction, entrepreneurial training, job-readiness training, and adult education and literacy.[27] Veterans holding a Gold Card receive one-on-one counseling services and case management for up to six months. Individuals who qualify for intensive services can also receive comprehensive and specialized assessments of skills and service needs, individual employment plans, employment counseling and career planning, case management, and prevocational or pretraining short-term skill development services. Staff are provided to specifically focus on the VETS Jobs for Veterans State Grants (JVSG) program. The JVSG program is a $175 million effort to help veterans with employment counseling. JVSG funds cover the costs for two positions, the Disabled Veterans' Outreach Program (DVOP) specialists and the local veterans' employment representative (LVER). In addition to the core services provided by the AJCs, DVOP specialists develop expertise in labor market and employment services that are specifically relevant to disabled veterans, and LVERs directly contact busi-

---

[26] Discussion with federal program manager, April 29, 2015.

[27] For more information on DoL's AJCs, see American Job Centers, website, n.d.

nesses, federal agencies, and associations of contractors and employers to encourage the hiring and advancement of qualified veterans.[28]

Career OneStop

Purpose: The Veterans ReEmployment section of the Career OneStop website aims to serve as a one-stop site for employment, training, and financial help for veterans.

Oversight: The website is funded by DoL.

Population: Many of the resources are not explicitly designed for veterans and service members; however, one section of the website is labeled "Veterans ReEmployment" and does provide information specific to these populations.

Services: The Career OneStop website is linked to AJCs and contains a large number of resources, including a tool kit with detailed information about occupations, training, wages, and other employment-related information; links to local resources that provide employment assistance; tip sheets on all aspects of the hiring process; and a search engine to find nearby AJCs. The website also provides a toll-free help line that can be used to speak directly with a staff member about employment resources and services. The website includes a military-to-civilian job search engine, information about benefit programs, information about education and training opportunities, and links to programs that focus on veterans and service members.[29]

O*NET and My Next Move for Veterans

Purpose: O*NET and My Next Move for Veterans classify occupations and provide data on occupational requirements and outlook.

Oversight: O*NET and My Next Move for Veterans are sponsored by DoL, Employment and Training Administration, and developed by the National Center for O*NET Development.

Population: While the resources on the website are targeted to individuals with military experience, they can be accessed by anyone.

Services: The main O*NET website and My Next Move for Veterans website both provide access to a set of search engines that allow veterans, service members, and others in the public to explore careers of interest and identify civilian occupations that might be related to prior military duties. Specifically, one of the search engines allows service members to enter a branch of service and military occupation code or title, and the search results give a list of occupations. The sites provide brief descriptions of each occupation, including information about the knowledge, skills, and abilities required; education, personality, and technology requirements; and average salary and other indicators of job prospects. In addition to the search engine and descriptions of specific occupations, the My Next Move for Veterans website provides resources to support career exploration, including tips for writing résumés that highlight related civilian skills and the O*NET Interest Profiler, which provides individuals with guidance on possible careers based on their revealed interests on the assessment.[30]

---

[28] For more information on DoL's Gold Card Program, see "New Employment Initiatives for Veterans," web page, U.S. Department of Labor, n.d.

[29] For more information on DoL's Career OneStop program, see Career OneStop, website, n.d.

[30] For more information on DoL's O*Net program see O*Net Online, website, n.d. For more information on DoL's My Next Move for Veterans, see My Next Move for Veterans, website, n.d.

Registered Apprenticeship (Employment and Training Administration)

Purpose: The Registered Apprenticeship program provides opportunities for workers seeking high-skilled, high-paying jobs, and it helps to support employers seeking to build a high-quality workforce.

Oversight: DoL's Office of Apprenticeship works in conjunction with State Apprenticeship Agencies to administer the program nationally.

Population: The program is broadly available to all Americans and is not explicitly targeted to reserve component members.[31]

Services: Participants in the program receive a combination of structured learning and on-the-job training through training centers and colleges and on-the-job training from an assigned mentor. The program has opportunities in construction and manufacturing, health care, energy, and homeland security. Apprenticeships range from one to six years, but the majority are four years. Individuals in the Registered Apprenticeship program earn a wage for the work they provide (starting at approximately $15 an hour for most positions), and this wage increases over time as they become more proficient. GI Bill funds can be used to cover the costs associated with formal learning experiences. Upon completion of a Registered Apprenticeship program, participants receive an industry-issued, nationally recognized credential that certifies occupational proficiency. According to the program's website, more than 170,000 individuals entered the program in fiscal year 2014. The program also offers preapprenticeship programs for individuals to prepare for a Registered Apprenticeship by providing basic information and training.[32]

Veterans' Employment and Training Service

Purpose: The VETS office is designed to prepare individuals for meaningful careers, provide employment resources and expertise, and protect people's employment rights.

Oversight: The office is overseen by DoL's Office of the Assistant Secretary for Veterans' Employment and Training Service.

Population: The VETS office serves veterans, transitioning service members, and their spouses. The office serves reserve component members but is not explicitly targeted to the reserve components.

Services: The office has national and regional offices that are staffed with individuals who can provide information about VETS programs and services. In addition, the office provides employment placement assistance through a network of AJCs and the Career OneStop website. The VETS office is also responsible for managing the DoL aspects of the Transition Assistance Program, including the Employment Workshop. The Employment Workshop covers such topics as career exploration and validation, job search planning, building an effective résumé, federal hiring, and interview skills. The VETS office also oversees the My Next Move for Veterans program (described above). The VETS website offers a variety of resources, including information about the Federal Veterans Hiring Initiative, information for reserve component members on the Uniformed Services Employment and Reemployment Rights Act (USERRA),

---

[31] While Registered Apprenticeships are broadly available to the public and not restricted to veterans or service members, the website does highlight several nonprofit programs that explicitly target veterans, including Helmets to Hardhats, Veterans in Piping, and the Painters and Allied Trades Veterans Program.

[32] For more information on DoL's Registered Apprenticeship program, see "Registered Apprenticeship," web page, U.S. Department of Labor, n.d.

access to fact sheets, and links to a range of programs and resources offered by DoL and other organizations. The website also provides information about the competitive grant opportunities that are available through VETS.[33]

Department of Veterans Affairs

The VA also offers a set of job placement and related employment programs for reserve component members. Most of these VA programs are targeted toward veterans, though many of them also allow for some use of services by transitioning service members, families, and caregivers. In some cases, the programs are specifically targeted to veterans with service-related injuries. None of the programs explicitly targets services to reserve component members, but if reserve component members are veterans, they might be eligible for some of the programs. We describe each of these programs and resources below.

Compensated Work Therapy
Purpose: Compensated Work Therapy (CWT) was developed to ensure realistic and meaningful vocational opportunities and successful reintegration into the community.

Oversight: CWT is administered by the Veterans Health Administration.

Population: CWT is targeted to honorably discharged veterans with physical and mental disabilities. While the program serves reserve component members who are qualifying veterans, the services are not explicitly targeted to the reserve components.

Services: CWT is composed of several different subprograms:

- Incentive Therapy (IT) is a program that provides work experience in the VA medical center for veterans who exhibit severe mental illness or physical impairments. Each participant in the IT program has an individual treatment or service plan, case manager, and treatment team to monitor services.
- Sheltered Workshop (SW) is a four- to six-month program that uses a simulated work environment to prepare and assess individuals for employment.
- Transitional Work (TW) is a preemployment vocational assessment program where participants are screened by vocational rehabilitation staff, assessed for ability, and matched to a work assignment. VA vocational specialists create individual treatment plans and oversee the progress of participants.
- The Supported Employment (SE) program was designed to help veterans with psychosis and other serious mental illness gain access to meaningful competitive employment. Veterans in the program engage in full- and part-time employment with appropriate supports and workplace accommodation, and supports are gradually phased out as individuals demonstrate the ability to work independently.
- The Transitional Residence (TR) program provides a rehabilitation-focused residential setting for veterans recovering from chronic mental illness, substance abuse, and homelessness. The program relies on a residential therapeutic community of peer and profes-

---

[33] For more information on DoL's VETS agency, see Veterans Employment and Training Service, website, U.S. Department of Labor, n.d.

sional support, and the goal is to increase personal responsibility and encourage individuals to achieve individualized rehabilitation goals.[34]

Education and Career Counseling and CareerScope

Purpose: The Education and Career Counseling program provides free education and career counseling for transitioning service members.

Oversight: The program is overseen by the VA, though the CareerScope tool is provided by a nongovernmental organization, the Vocational Research Institute.

Population: The program is available to transitioning service members, including individuals transitioning out of active duty (from six months before to one year after leaving service), individuals who are beneficiaries of or qualify for certain military education benefit programs, and veterans and their dependents who qualify for those military education benefit programs.[35] Members of the reserve components who are in the Selected Reserve, have been called to active duty, or otherwise qualify for military education benefits are eligible under these requirements.

Services: One-on-one counseling services might involve career decisionmaking for civilian or military occupations, educational and career counseling to choose an appropriate civilian occupation and develop a training program, and academic and adjustment counseling to resolve barriers that impede success in training or employment. The VA also directs individuals to a tool called CareerScope, which provides an assessment of interests and aptitudes, gives recommendations about which careers an individual might enjoy and be successful doing, and provides information about the courses or training programs to focus on to pursue those careers. The VA covers the cost of assessment for all eligible recipients.[36]

Office of Small and Disadvantaged Business Utilization

Purpose: The Office of Small and Disadvantaged Business Utilization (OSDBU) was designed to connect veteran entrepreneurs to relevant best practices and information.

Oversight: The OSDBU is a VA office that administers the Veterans Entrepreneur Portal (VEP) and the Vets First Verification program.

Population: All of the information on the website can be accessed by the general public; while some of the information is specific to veteran entrepreneurs, some of the information is also general business guidance. The Vets First Verification Program provides services to veterans and their surviving spouses, so reserve component members who are veterans qualify for those programs.

Services: The OSDBU website provides the VEP, which provides information about starting and growing a business, accessing financing, dealing with federal agencies, and a range of other topics relevant to entrepreneurs. In addition to the VEP, OSDBU administers the Vets First Verification Program. Businesses must first be verified as a Service-Disabled Veteran-Owned Small Business/Veteran-Owned Small Business (SDVOSB/VOSB). To support

---

[34] For more information on VA's CWT program, see "Compensated Work Therapy," web page, U.S. Department of Veterans Affairs, n.d.

[35] The education programs that determine eligibility for non–active duty personnel include the Montgomery GI Bill, the post-9/11 GI Bill, Vocational Rehabilitation and Employment, the Veterans Education Assistance Program, Dependent's Education Assistance, the Montgomery GI Selected Reserve program, and the Reserve Educational Assistance Program.

[36] For more information on CareerScope, see "CareerScope," web page, U.S. Department of Veterans Affairs, n.d.

the verification process, the program provides online informational resources and access to counselors to help veteran entrepreneurs through the process. As soon as a business is verified, it qualifies for special set-asides, along with women-owned small businesses and other types of businesses that are given preferences in contracting with the VA for the provision of goods and services. The highest preference is given to SDVOSB businesses, followed by VOSB businesses.[37]

VA for Vets

Purpose: VA for Vets is designed to attract, retain, and support individuals in careers at the VA.

Oversight: VA for Vets is a program under the VA's Veteran Employment Services Office (VESO).

Population: Most resources are designed for all veterans and transitioning service members and are not exclusively targeted toward reserve component members. However, according to the program website, the resources for current employees explain that "VA for Vets offers resources and support so that you can continue your VA career, while serving in the National Guard or Reserve," indicating that the program does intend to serve members of the reserve components.

Services: The program's website provides a range of informational resources, including fact sheets, frequently asked questions, videos, and links to other employment program resources. Among the materials provided on the site, the resources for prospective employees provide information about rules and regulations around VA employment (e.g., veterans' preference, OPM regulations) and guidance about the hiring process and activities after applying. Resources for current employees provide information about mentorship opportunities, supervisor guidelines, and general employment issues. The materials for human resource professionals and supervisors cover similar topics. In addition to the online resources, the program provides access to regional veteran employment coordinators through a call line. These individuals are available to provide more information to veterans and transitioning service members about the opportunities at the VA, as well as offering to facilitate the hiring process through one-on-one assistance to human resource professionals with open positions.[38]

Veterans Employment Center

Purpose: The VEC is the federal government's main online portal for connecting transitioning service members, veterans, and their families to public- and private-sector career opportunities.

Oversight: The VEC is overseen by the VA.

Population: The eBenefits website on the VEC was designed to serve transitioning service members, such as veterans (including wounded warriors), service members, their family members, and their authorized caregivers. All of the resources provided by the VEC are available to reserve component members, though the VEC was not designed to exclusively target the reserve components.

Services: The VEC provides information and services on benefits across all areas, including health, education, and employment. The VEC focuses on providing access through the website to tools and informational resources that can assist individuals with job searches and

---

[37] For more information on VA's Office of Small and Disadvantaged Business Utilization, see "Office of Small and Disadvantaged Business Utilization," web page, U.S. Department of Veterans Affairs, n.d.

[38] For more information on the VA for Vets program, see VA for Vets, website, U.S. Department of Veterans Affairs, n.d.

assist employers with hiring. The primary tool offered through the VEC is a job board that connects individuals looking for jobs to employers that are committed to hiring veterans, service members, and their families. Other tools provided by the VEC include a skill translator to identify civilian skills that correspond with those gained through military experience, as well as a profile and résumé-builder that assists individuals in developing their résumés and posting them for employers to access. In addition to these tools, the VEC provides links to descriptions of other programs and resources, including resources in the areas of employment, education, training programs, and disability. Employers are able to access resources that provide information about the military experience and offer advice about supporting veterans, service members, and their families in the workplace. The VEC also provides information about job fairs for individuals and employers.[39]

Veterans Opportunity to Work

Purpose: The Veterans Opportunity to Work (VOW) program was designed to provide an extension of the vocational rehabilitation services provided by the Vocational Rehabilitation and Employment (VR&E) program.

Oversight: The program is administered by the VA's Veterans Benefits Administration.

Population: Individuals are eligible for extended, individualized employment assistance if they completed VocRehab under the VR&E program and apply within six months of using an initial claim for unemployment benefits. The requirement for prior receipt of VR&E services means that only active duty service members and veterans with disabilities are eligible for the extended employment services.

Services: The VOW program also provides a Special Employer Incentives (SEI) program that provides financial incentives to employers to encourage them to hire veterans who face extraordinary barriers to employment. According to the website, employers are reimbursed for up to half the veteran's salary, to cover certain supplies and equipment, additional instruction expenses, and any loss of production. VA counselors under the SEI program provide individualized assistance to employers to prepare paperwork and match employers with service-disabled veterans. In addition to the rehabilitation and SEI services, VOW is responsible for the VA's component of the Transition Assistance Program, which is described below. Historically, the VOW program also oversaw the Veterans Retraining Assistance Program (VRAP), which provided training benefits to unemployed veterans over the age of 35. However, the VRAP program ended in March 2014. As of July 2015, the VOW website continued to provide information about VRAP but noted that it had ended. In addition to describing VOW's major programs, the VOW website provides some web-based informational materials, such as a list of high-growth occupations and links to other VA employment programs and resources described in this chapter.[40]

VetSuccess on Campus

Purpose: The VetSuccess on Campus (VSOC) program assists student veterans and their qualified dependents through the coordinated delivery of on-campus benefits assistance and coun-

---

[39] For more information on the VEC, see "Job Seekers," job bank, Veterans Employment Center, eBenefits, n.d.

[40] For more information on the VA's VOW program, see "Veterans Opportunity to Work," web page, U.S. Department of Veterans Affairs, n.d.

seling intended to support the successful completion of education and preparation to enter the labor market.

Oversight: The VSOC program appears to be run by the VR&E program.

Population: To qualify for the individualized counseling provided by VSOC, transitioning service members must be within six months of discharge from active duty, and veterans must be within one year of discharge from active duty. Student veterans who do not fall within these categories are also covered under the program if they are receiving education assistance from the GI Bill programs, VR&E, the Reserve Educational Assistance Program, Post-Vietnam Era Veterans Educational Assistance Program, or the Dependents' Educational Assistance Program. VSOC counselors provide a range of services, including community and on-campus outreach, educational and vocational assessments and counseling, and referrals to other programs and services. As of fiscal year 2013, the VSOC program was available on 94 college campuses.[41]

Vocational Rehabilitation and Employment

Purpose: The VR&E program was designed to help individuals access and participate in community resources and enhance independence in daily life.

Oversight: The VR&E program is administered by the VA's Veterans Benefits Administration.

Population: While the website and web-based resources are available to the general public, the vocational rehabilitation (referred to as VocRehab) services are restricted to current transitioning active duty service members and veterans with disabilities, and the VR&E website notes that the program was designed to target these individuals. To qualify for individualized VR&E services, individuals are also required to submit an application and must not have had a dishonorable discharge. Among the reserve components, only disabled veterans qualify for this program.

Services: The program provides individualized employment services to help with job training, employment accommodations, résumé development, and job search coaching. The VocRehab program offers a range of services, including vocational counseling, résumé development, assistance getting a job, apprenticeships and internships, and assistance with other types of rehabilitation services (e.g., education, medical care). Individuals are connected directly with vocational rehabilitation counselors and ECs through one of several tracks, including reemployment, rapid access to employment, self-employment, employment through long-term services, and independent living. In addition to resources on VocRehab, the VR&E website provides links to several other VA employment programs and resources, such as the VOW program. The website includes web-based guidance documents, such as tips on employment for veterans and military-to-civilian skill-translation advice.[42]

Workstudy Program

Purpose: The Workstudy Program provides individuals enrolled in postsecondary education with work-study opportunities.

---

[41] For more information on VA's VSOC program, see "VetSuccess on Campus," web page, U.S. Department of Veterans Affairs, n.d.

[42] For more information on the VA's Vocational Rehabilitation and Employment program, see "Vocational Rehabilitation and Employment (VR&E)," web page, U.S. Department of Veterans Affairs, n.d.

Oversight: The program is administered by the VA's Veterans Benefits Administration.

Population: The Workstudy Program is available to service members, veterans, and family members who are enrolled in college at least 75 percent of the time under qualifying military education programs (e.g., GI Bill, Reserve Education Assistance Program). Reserve component members who meet these conditions are eligible for this program, though the program is not explicitly targeted to these individuals.

Services: The program provides funding for work at an hourly wage equal to the federal or state minimum wage (the state minimum wage is used if it exceeds the federal minimum wage). The program allows for the work to extend up to 25 weeks per year of enrollment and covers employment at higher-education institutions, DoD, and the VA. Individuals must apply and be matched with a qualifying work-study position.[43]

Interagency Programs and Other Federal Programs

There are several programs offered jointly by several federal agencies, or offered by federal agencies outside of DoD, DoL, and the VA. We describe these programs and resources below.

Feds Hire Vets

Purpose: Feds Hire Vets provides consistent and accurate information, useful training, and other resources to individuals.

Oversight: Feds Hire Vets acts as OPM's primary governmentwide veterans' employment website. The website was developed in response to the 2009 executive order that also established Veterans Employment Program Offices (VEPOs) in each of the 24 federal agencies.[44]

Population: The program serves veterans, transitioning service members, their families, federal human resources professionals, and hiring managers. The website has separate pages for veterans, transitioning service members, and family members.

Services: The website includes a variety of resources, including information sheets about preferences and hiring authorities, training modules on the federal hiring process, a link to USAJobs (the online portal for all federal applicants) to search available positions, and links to other potentially useful resources on federal hiring. The veteran employees' site includes a special section that specifically highlights information that reserve component members might need to know, including explanations of USERRA, reservist differential pay, military leave, and leave without pay. Information for hiring managers includes training for human resources practitioners and fact sheets on the value of hiring veterans and hiring authorities. The website also contains success stories and frequently asked questions for individuals interested in federal employment.[45]

---

[43] For more information, see "Workstudy," web page, U.S. Department of Veterans Affairs, n.d.

[44] See Executive Order 13518, *Employment of Veterans in the Federal Government*, Washington, D.C.: The White House, Office of the Press Secretary, November 9, 2009.

[45] For more information on OPM's Feds Hire Vets webpage, see Feds Hire Vets, website, n.d.

National Resource Directory

Purpose: The National Resource Directory now focuses exclusively on providing a searchable database that allows individuals to find federal, state, and local resources that are available to assist with various benefit areas, including employment.

Oversight: The National Resource Directory is housed on the VA's eBenefits website and is overseen by DoD, DoL, and the VA.[46]

Population: The National Resource Directory is targeted to all service members, veterans, and their families, and some of the resources listed in the National Resource Directory are explicitly designed for reserve component members.

Services: As of April 2015, the National Resource Directory listed 1,402 different resources available for employment assistance, including job application sites for organizations, such as the Alaska Department of Motor Vehicles and the Seattle-Tacoma International Airport; informational sites about employment for individuals with posttraumatic stress disorder and other disabilities that might affect employment prospects; and nonprofit service providers that offer support to individuals seeking employment. Resources are also provided by special category in three areas: (1) resources for employers and veterans' employment service providers; (2) specialized support and resources for wounded warriors; and (3) transitioning from a military to a civilian career. The resource summaries contained in the directory include a one-sentence description, a web link for the program or resource, and an opportunity for users to provide feedback on the program or resource with a "thumbs up" or "thumbs down." On May 20, 2015, a search of the term *reserve* found 92 resources on employment benefits, including governmental and nongovernmental resources, and a search of *National Guard* found 91 resources on employment.[47]

Transition Assistance Program

Purpose: In 2013, TAP was redesigned through an interagency collaboration between DoD, DoL, the VA, the Small Business Administration, Department of Education, and OPM to assist transitioning service members.

Oversight: TAP is overseen by DoD, DoL, VA, the Small Business Administration, Department of Education, and OPM.

Population: Under Title 10, all transitioning service members are required to take part in the program—including guard and reserve members demobilizing after 180 days or more of active service. The TAP Virtual Curriculum is available to nonveteran reserve component members (those who were not deployed for 180 days or longer), and reserve component members can also attend the brick-and-mortar TAP courses if there is availability, but TAP courses are designed primarily for active duty service members and reserve component members who have been activated for at least 180 days.

Services: The program for service members is referred to as Transition GPS. Services provided under Transition GPS include briefings on VA benefits, individual transition planning, employment workshops, and tailored tracks. These tracks provide information that is targeted to individuals who are focused on a particular area of development and include a technical-career training track, an education track, and an entrepreneurship track. Successful comple-

---

[46] The VA's eBenefits website is the VA's main portal to information regarding VA benefits.

[47] For more information on the National Resource Directory, see National Resource Directory, web page, eBenefits, Department of Veterans Affairs and the Department of Defense, n.d.

tion of the program involves participation in three components: preseparation counseling to develop an individual transition plan and identify career planning needs; VA Benefits Briefings I and II, which explain what benefits the service member has earned and how to obtain them; and the DoL's Employment Workshop, which focuses on the specifics of obtaining employment in today's job market, including résumé preparation and interview practice. Individuals are assessed by commanders to determine whether they meet career-readiness standards no fewer than 90 days prior to separation. If a service member has not met standards or does not have a viable transition plan, the commander will put the service member directly in contact with a partner agency that will provide continued support. Beginning in 2015, TAP will be provided throughout a service member's time in the military, according to his or her Military Life Cycle. This proactive approach will help service members plan ahead to bridge their military and civilian careers by providing time and resources to conduct career planning activities during key touch points in their military service.[48]

The Veterans Business Outreach Program

Purpose: The mission of the Veterans Business Outreach Program (VBOP) is to maximize the availability, applicability, and usability of small-business programs.

Oversight: VBOP is housed in the Small Business Administration agency, under the Office of Veterans Business Development.

Population: Veterans, service-disabled veterans, reserve component members, and their dependents or survivors are eligible for the program.

Services: The VBOP program is designed to provide entrepreneurial development services, including business training, counseling and mentoring, and referrals for veterans who own or are considering starting a small business. The program contracts with 15 service providers across the county to serve as Veterans Business Outreach Centers (VBOCs). VBOCs provide a range of services, including assessment of entrepreneurial needs and requirements, assistance designing a five-year business plan, assistance in assessing the strengths and weaknesses of the business plan, targeted training and counseling to service-disabled veterans, on-site visits to business, review of financial statements, and guidance on other such business-related activities as franchising and marketing. In addition to describing these services, the VBOP website provides a list of links to additional governmental and nonprofit programs and resources.[49]

Veterans Employment Program Offices

Purpose: On November 9, 2009, President Barack Obama signed Executive Order 13518, *Employment of Veterans in the Federal Government*, which established the Veterans Employment Initiative. The initiative is a strategic approach to helping veterans, service members, and their spouses to find employment in the federal government.

---

[48] DoL's portion of the five-day TAP curriculum is a three-day employment workshop at military installations worldwide for thousands of separating service members. For veterans and military spouses who might also benefit, the full TAP curriculum is available online. DoD's one-day portion of the TAP curriculum focuses on financial management and personal resilience. VA's one-day portion focuses on the benefits that transitioning service members are eligible to receive. For more information on TAP, see "Transition Assistance Program," web page, U.S. Department of Labor, n.d.; Transition Assistance Program, website, U.S. Department of Defense, n.d.; "Transition Assistance Program," web page, U.S. Department of Veterans Affairs, n.d.

[49] For more information on the VBOP program, see "Veterans Business Outreach Program," web page, U.S. Small Business Administration, n.d.

Oversight: Each of the 24 federal agencies has created a VEPO.

Population: The program resources are available broadly to veterans, transitioning service members, and spouses. The programs are required to report quarterly to OPM on the number of individuals who served in particular subgroups, including disabled veterans, women veterans, and spouses of active duty service members. While reserve component members are not specifically targeted by the program, they should qualify for all of the services provided.

Services: VEPOs and their staffs provide advice and guidance to veterans and their spouses, and some offices focus particularly on severely injured veterans. The websites for the various VEPO programs provide a range of resources, including information about hiring authorities, links to other programs and resources (e.g., USAJobs, VA for Vets), and a description of the program's services. Some of the VEPO offices, such as the one housed at the U.S. Department of Agriculture, provide one-on-one services to individuals, including information sessions on the agency and hiring processes, assistance with résumés, and referrals for open positions. Other VEPO programs appear to focus on providing online resources to individuals through a dedicated website.[50]

Summary of Findings

In total, we found 40 federal programs, resources, and offices that provide job placement assistance to reserve component members. The bulk of the job placement assistance and related employment services available to reserve component members are located within DoD, DoL, and the VA. As discussed in the next chapter, it is important to note that many of the programs (including most of the VA programs) are targeted to a smaller population (e.g., service-disabled veterans), of which only a small portion of the reserve component is eligible.

---

[50] For more information on VEPOs, see Executive Order 13518, 2009.

CHAPTER THREE

Gaps and Overlaps in Federal Job Placement Assistance for Reserve Component Members

A comprehensive gap analysis would require a comprehensive assessment of the employment needs of reserve component members, as well as evaluations of the programs that provide employment services to reserve component members. The study team could not find a comprehensive assessment that identifies the employment needs of reserve component members, and our sponsor was not aware of such a needs assessment. Therefore, to assess potential gaps in federal employment programs, we developed a framework for assessing the programs' services against the needs that reserve component members have expressed in previous work we have conducted on this topic (e.g., the need for assistance with legal issues), as well as typical employment-related services for the general public.[1] These categories of services include

- assistance with résumé and interview preparation
- assistance with job search and placement
- assistance with job and career planning
- skills assessment and certification
- hosting job fairs
- assistance with legal issues
- assistance with entrepreneurship
- internships, apprenticeship, and training
- financial assistance
- access to information and tools.

We then accounted for whether a program, resource, or office offers each of these ten categories of services, and during our informational discussions, we also asked federal program managers whether there are any gaps or overlaps in federal employment services for reserve component members. All of this information was then compiled into a matrix that allowed us to visually identify gaps and overlaps (see Table 3.1). In addition to accounting for the services provided by each program, resource, or office, we accounted for the intensity of services. We define *high-touch* programs as those that provide opportunities for extended follow-on services that are typically personalized to meet individual needs. We define *low-touch* programs and resources as those that provide one-time services for a short, those that do not provide personalized services, and those that do not provide any human component.

---

[1] For perspectives on reserve component employment needs, see Werber, Schaefer, et al., 2013; also see Werber, Wenger, et al., 2015. The study team also identified typical employment services offered by private employment agencies and state employment offices.

Overview of Services Provided

We identify several patterns in Table 3.1. The most commonly offered services include access to information and tools, assistance with job search and placement, and assistance with résumé and interview preparation. The programs, resources, and offices that provide access to information and tools can be found across agencies and are a common component of both high-touch and low-touch programs. More than half of the programs, resources, and offices that provide assistance with job search and placement and assistance with résumé and interview preparation are located in DoD, and these programs, resources, and offices are typically high-touch in nature. In general, high-touch programs, resources, and offices are more likely to be found in DoD.

It is important to note that while some of the services are less common, it does not necessarily imply a gap in services. The more commonly provided services might be those that are most needed by service members, or those services that are most effective in securing employment.

The less commonly provided services might be those that serve a unique need held by a smaller subset of service members. Alternatively, we do not necessarily interpret services provided by a large number of programs to represent overlap. If these programs are targeting different populations or are providing services in a unique way that serves the needs of particular service members, the duplication of services might be optimal. In the following sections, we describe what we do see as potential areas of gaps and overlaps.

Potential Gaps in Services Provided

Without a comprehensive needs assessment, it is unclear whether all needs are being met. However, we found that the programs, resources, and offices we identified provide a wide range of different high-touch and low-touch services, and we did not identify any obvious gaps in services in the current portfolio of federal job placement assistance and related employment programs available to reserve component members. However, our discussions with key stakeholders suggest that there are some potential areas of concern if programs are consolidated. Based on these discussions and prior studies on services for veterans and service members, federal agencies should consider maintaining the following types of programs to avoid future gaps in services:

- *High-touch programs*: There is evidence from prior research[2] and our discussions with stakeholders that high-touch services are critical to retain. Programs with high-touch approaches allow assistance to be tailored to unique individual needs and allow for a range of benefits beyond basic employment services, including a feeling by service members that they are supported and understood. While high-tech tools, databases, and high-quality informational websites can act as important resources for individuals and service providers, they cannot replace the services provided by these high-touch programs. For example, a job counselor can follow an individual through every aspect of the job search, including offering referrals to technological resources that can help to meet needs

---

[2] See Werber, Wenger, et al., 2015.

Table 3.1
Services Provided and Populations Served by Federal Job Placement Assistance Programs and Resources

Agency	Program Name	Population Served	Targeted to Reserve Component	Assistance with Résumé and Interview Prep	Assistance with Job Search and Placement	Assistance with Job and Career Planning	Skills Assessment and Certification	Hosting Job Fairs	Assistance with Legal Issues	Assistance with Entrepreneurship	Assistance with Internships, Apprenticeships, and Training	Financial Assistance	Access to Information and Tools
DoD	Airman and Family Readiness	Air Force RC members	No	X	X								X
	Always a Soldier (AAS)	RC members with service-related injuries	No		X						X		
	Beyond Yellow Ribbon (BYR) Programs	All RC members	Yes	X				X	X				X
	Career Skills Program (CSP)	Activated Army RC members	No				X					X	X
	Credentialing Opportunities Online (COOL)	All RC members	No								X	X	X
	DoD Hiring Heroes	RC members with service-related injuries	No	X	X	X	X						
	Education and Employment Initiative	RC members with service-related injuries	No			X	X	X					X
	Employment Readiness Program (ERP)	Army RC members	No	X	X					X			X
	Fleet and Family Support Centers	Navy RC members	No	X	X					X			X
	Guard Apprenticeship Program Initiative (GAPI)	Army RC members	Yes								X		
	Hero2Hired (H2H)	All RC members	Yes		X			X					X
	Job Connection Education Program (JCEP)	Army RC members in four states	Yes	X	X	X							X

Table 3.1—Continued

Agency	Program Name	Population Served	Targeted to Reserve Component	Assistance with Résumé and Interview Prep	Assistance with Job Search and Placement	Assistance with Job and Career Planning	Skills Assessment and Certification	Hosting Job Fairs	Assistance with Legal Issues	Assistance with Entrepreneurship	Assistance with Internships, Apprenticeships, and Training	Financial Assistance	Access to Information and Tools
DoD	Marine and Family Program's Career Services	Marine RC members	No	■	■	■							■
	Military OneSource	All RC members	No	■	■	■							■
	National Guard Employment Network (NGEN)	All RC members	Yes	■	■	■	■						■
	Operation Warfighter (OWF)	RC members with service-related injuries	No		■						■		
	Partnership for Youth Success (PaYS)	Army RC members	No			■							
	Public Private Partnership (P3) Program	Army RC members	Yes	■	■						■		■
	SkillBridge	Transitioning RC members	No										
	Troops to Teachers (TTT)	Transitioning RC members	No				■						■
	Yellow Ribbon Reintegration Program (YRRP)	Activated RC members	Yes					■	■				■
DoL	American Job Centers (AJCs)	All RC members; veterans receive additional services	No	■	■	■	■	■		■	■		■
	Career OneStop	All RC members	No		■								■
	O*NET and My Next Move—For Veterans	All RC members	No	■			■						■

Table 3.1—Continued

Agency	Program Name	Population Served	Targeted to Reserve Component	Access to Information and Tools	Financial Assistance	Assistance with Internships, Apprenticeships, and Training	Assistance with Entrepreneurship	Assistance with Legal Issues	Hosting Job Fairs	Skills Assessment and Certification	Assistance with Job and Career Planning	Assistance with Job Search and Placement	Assistance with Résumé and Interview Prep
DoL	Registered Apprenticeship	All RC members	No	■		■							
	Veterans' Employment and Training Service (VETS)	All RC members	No		■						■	■	■
VA	Compensated Work Therapy (CWT)	RC members with service-related injuries	No			■				■		■	
	Education and Career Counseling and CareerScope	Transitioning RC members; RC members qualifying for educational benefits	No							■	■		
	Office of Small and Disadvantaged Business Utilization (OSDBU)	Veteran RC members	No	■	■	■	■						
	VA for Vets	Veteran RC members	No	■				■					
	Veterans Employment Center (VEC)	All RC members	No	■									
	Veteran's Opportunity to Work (VOW)	Veteran RC members completing VocRehab	No	■	■							■	
	VetSuccess on Campus (VSOC)	Veteran RC members; RC members qualifying for educational benefits	No							■			
	Vocational Rehabilitation and Employment (VR&E)	RC members with service-related injuries	No							■			

Table 3.1—Continued

Agency	Program Name	Population Served	Targeted to Reserve Component	Assistance with Résumé and Interview Prep	Assistance with Job Search and Placement	Assistance with Job and Career Planning	Skills Assessment and Certification	Hosting Job Fairs	Assistance with Legal Issues	Assistance with Entrepreneurship	Assistance with Internships, Apprenticeships, and Training	Financial Assistance	Access to Information and Tools
VA	Workstudy Program	RC members qualifying for educational benefits	No		■							■	
Other federal agencies and joint programs	Feds Hire Vets	Veteran RC members	No						■		■		■
	National Resource Directory	All RC members	No										■
	Transition Assistance Program (TAP)	Activated RC members	No	■		■				■			■
	Veterans Business Outreach Program (VBOP)	All RC members	No								■		■
	Veterans Employment Program Offices (VEPOs)	Veteran RC members	No	■	■								■
Number of programs, resources, and offices offering services				16	23	15	11	6	4	6	13	6	29

NOTES: Programs in bold are ones that provide high-touch services. RC = reserve component.

(e.g., skill translators, the VEC job portal); providing in-person assistance with needs that cannot be addressed with technology (e.g., building confidence, developing individualized career plans); and acting as an advocate for service members when engaging with employers.

- *Strong connections to employers*: It is critical that federal employment programs prioritize focusing on strong connections to employers. Prior research[3] and our discussions with stakeholders indicate that the effectiveness of employment programs can be greatly enhanced if these programs are also seen to meet employer needs. Employers have many job boards where they can post positions, and there are many applicants to choose from. To ensure that veterans and service members are given priority, employers must be informed about the benefits of hiring these individuals, and employers must feel that investing time with employment programs provides a return to their companies.

- *Data-collection and case-management capabilities*: One aspect of many high-touch programs is the ability to track data, including individual contact information, progress on various aspects of the job search, and eventual employment outcomes. Several government programs do not collect data that differentiate veteran versus nonveteran active duty, reserve, and National Guard status. Collecting and reporting program data to more specifically differentiate veterans from nonveterans, and to distinguish active duty from reserve and National Guard status, should be a priority in the future. Case-management data can also help programs efficiently and effectively deliver high-touch services to service members, allowing counselors to follow up with individuals and tailor services to their needs. In addition, programs can use the data to follow up with employers to gather feedback and hold employers accountable for their commitments to hire service members and veterans. Data on activities and outcomes are also important in assessing how program resources are being used and whether programs are meeting their intended goals.

Navigational Barriers to Accessing Services

Given the large number of federal employment programs and resources offered by different agencies and designed to serve different populations, one major finding that emerged from our search is that it is extremely difficult to navigate the large number of programs and resources available to service members. Other studies of programs for service members argue that overlap and a lack of coordination can make navigation difficult.[4] It was even difficult for our research team to tease out the relationships and differences between some programs and resources. Also, when we presented our list of programs and resources to one of the stakeholders for his feedback, he commented that if he were a service member, he would not know where to start if he were in need of job placement assistance. A potential gap, therefore, is a lack of clear guidance on how to navigate this complex web of programs.

There are some programs, resources, and offices that attempt to guide service members to federal employment-related services. For example, such websites as the National Resource

---

[3] See Kimberly Curry Hall, Margaret C. Harrell, Barbara A. Bicksler, Robert Stewart, and Michael P. Fisher, *Veteran Employment: Lessons from the 100,000 Jobs Mission*, Santa Monica, Calif.: RAND Corporation, RR-836-JPMCF, 2014; Werber, Wenger, et al., 2015.

[4] U.S. Government Accountability Office, 2015; Ellen M. Pint, Amy Richardson, Bryan W. Hallmark, Scott Epstein, and Albert L. Benson, *Employer Partnership Program Analysis of Alternatives*, Santa Monica, Calif.: RAND Corporation, TR-1005-A, 2012.

Directory, Military OneSource, and Career OneStop attempt to provide information about many of the federal programs that are available. In fact, most of the websites we visited provided links to other programs or resources. However, these websites and search tools typically only capture a small portion of the overall federal portfolio of job placement assistance and related employment programs. And, for the most part, the information provided about programs and resources in these search tools are insufficient for a service member to quickly identify which of the many programs and resources best meet a particular need.

In addition to these web resources, the TAP program offers workshops to provide individuals with information about employment programs and resources. However, these services are offered only on return from deployment and are therefore one-time services offered to a subset of the reserve components. Universal access to complete and updated information on federal programs and guidance for unique individual needs could be of great assistance to reserve component members.

Another effort that would simplify the navigation process would be to consolidate programs and resources to minimize the total number of programs and resources available. We discuss our recommendation related to this in Chapter Four. But even with a smaller federal portfolio, it might still be a challenge for reserve component members to identify the appropriate programs and services, given variation in services and eligible populations, and there might also be confusion among federal, state, and nongovernmental programs and resources.

Potential Overlaps in Services Provided

While we did not find any significant gaps among the many federal programs and resources that are available, we did find some potential areas of overlap. We describe below the three areas of overlap we identified: (1) many high-touch programs, (2) duplication of resources and tools, and (3) overlaps with other governmental and nongovernmental job placement assistance.

Many High-Touch Programs

We described the importance of having high-touch employment programs to allow for individualized assistance to service members and employers. However, we did find a large number of high-touch programs that are providing relatively similar services, including help with résumés, interviews, and job searches. In many cases, these programs define eligibility differently and target services to different populations, and it might be necessary to have some overlaps in programs to meet the unique needs of different populations. And, in some cases, these programs might offer other services that are distinct from other overlapping programs. In addition, several of the stakeholders we spoke with noted that smaller, decentralized programs were better able to develop the relationships with service members and employers that are necessary to support effective high-touch employment programs. However, these potential benefits of overlapping programs must be balanced against the potential costs. In addition to the resources that are required to support overlapping programs, a large number of programs that offer similar services with slight differences in approaches and populations served can be confusing to service members and employers and might affect the ability to place individuals into jobs.

Duplication of Resources and Tools

In addition to a large number of high-touch programs, there are many resources and tools that have been developed to support service members, and, in many cases, these resources and tools are duplicative. For example, we identified at least five different military-to-civilian skill translators referenced on federal employment program websites. Many programs also advertise skill translation as a key service that employment counselors offer to service members. In addition to these skill translators offered by federal programs and websites, there are many others offered from nongovernmental sources (e.g., websites for Boeing and Lockheed Martin). It is unclear why several different skill translators are needed, and multiple skill translators might create confusion among service members and providers of employment services. In addition to overlaps in skill translators, several different programs offer services, courses, or tip sheets to assist with basic employment-related tasks, such as résumé development and interviewing. There might be some need to tailor advice on these basic tasks based on individual characteristics and background and variation in employers' needs and preferences. However, there might be ways to streamline and align these services and resources to save time and money and to ensure that service members are being provided with consistent guidance and information across programs and resources.

Overlap with Other Governmental and Nongovernmental Job Placement Assistance

This study was commissioned to specifically examine the federal portfolio of employment programs and resources, so efforts to examine state and local programs and nongovernmental programs were beyond the scope of the study. However, it is critical to acknowledge that there is significant overlap between federal programs and resources and what is offered by nonfederal programs and resources. For example, a May 20, 2015, Google search of *employment assistance for veterans* found 131,000,000 results. The first page of results found three federal programs or resources, including VETS, VA Jobs, and VA Careers. However, it also listed a state program in California and four nongovernmental resources, including Hiring Heroes USA, Vet Jobs, and Veterans Inc. Efforts to streamline any overlaps in federal programs will certainly be helpful in addressing some of the confusion that service members might face when trying to sort through the large pool of programs and resources, but overlap with state, local, and nongovernmental programs will continue. This suggests another reason why service members might require guidance in sorting through the network of resources available to them.

Overview of Populations Served

After reviewing and describing the various federal programs offering job placement assistance and related employment services, we identified the reserve component members served by each program. We also determined whether the program explicitly targets services to reserve component members, as several of our interviewees noted the importance of programs that are designed for reserve component members, because they thought that these programs better accommodate the unique employment needs of reserve component members. The information about populations served by each program, resource, and office in our study is presented in Table 3.1.

As noted in Chapter Two, we found that DoD programs are typically offered to current and transitioning service members, and, in some cases, veterans are also eligible. DoL programs are typically available to all Americans, and some of the services are also targeted toward veterans. VA programs are typically targeted toward veterans, and while the programs typically include transitioning service members among eligible populations, few of the programs place an emphasis on providing outreach and services to reserve component personnel without veteran status. None of the programs, resources, or offices identified in our study outside DoD exclusively targets the reserve components.

Potential Gaps in Populations Served

Our review of the existing federal portfolio suggests that the large number of federal employment programs, resources, and offices available to reserve component members covers the full reserve component population (and many others as well). There are programs, resources, and offices that are available to reserve component veterans or other more-specific populations, and there are programs, resources, and offices that are broadly available to all members of the reserve components. Some programs target individuals with interests in specific occupations (e.g., teaching) and some programs even target entrepreneurs. Many of the programs and resources are designed to serve both employers and job seekers to ensure the targeting of all relevant stakeholders.

However, while many of these programs and resources are accessible to members of the reserve components, only seven of them are specifically targeted to and designed for the reserve components. All of the programs targeted to members of the reserve components are overseen by DoD. Most of the programs offered by DoL are broadly available to all Americans, and most of the VA programs have limited populations, such as veterans, veterans with service-related disabilities, and individuals utilizing educational benefit programs.

As described in Chapter One, members of the reserve components are likely to face different constraints from what faces individuals in the active component, spouses and families, and those in the general public who are seeking employment. To the degree that overlap in programs and resources might represent the targeting of services to different populations, this overlap might be providing an important benefit. However, it is important to consider whether there are low-cost ways of continuing to effectively meet the needs of reserve component members.

While we did not find any large gaps in the populations being served by federal employment programs, there are some populations that might not find the programs accessible, and our interviewees and prior research suggest that potential gaps might emerge if programs are consolidated. Below we describe four populations of reserve component members that might be particularly vulnerable to potential gaps in services:

- *Reserve component personnel without veteran status*: A large number of programs and resources are targeted to individuals who have been activated for at least 180 days (though many also report that they will broadly provide services to "transitioning service members"). To the degree that these programs and resources are intended to provide recognition and support for individuals who have been on active duty for at least six months, the targeting of programs to veterans might be appropriate. However, veterans account for less than half of the overall reserve components.[5] Employment programs that more broadly serve the full reserve component population might provide important benefits to recruitment, retention, and resilience. It is important to ensure that reserve component personnel without veteran status continue to receive access to programs and resources for job placement assistance and related employment services.

- *Young reserve component members*: We heard from several stakeholders that the youngest individuals in the reserve components might need job placement assistance and related employment services the most. These individuals often have little experience in the business world and lack familiarity with the job process. As indicated in Chapter One, younger members of the reserve components are more likely to be unemployed. However, younger members of the reserve components are often overlooked because they do not have veteran status and are more reluctant to look for help.[6] It is important for these young reserve component members to retain access to job placement assistance and related employment programs and services.

- *Underemployed individuals*: Many of the federal job placement assistance and related employment programs only accept unemployed service members or focus primarily on serving the unemployed. However, we heard from several stakeholders that underemployment is a major challenge that many in the reserve components face. Programs and resources that can help individuals find full-time employment to develop careers rather than jumping from job to job can have important impacts on the retention and resilience of reserve component members. Underemployed individuals should continue to have access to employment programs and resources, and federal agencies might want to consider how services can be better targeted to meet the needs of these individuals.

- *Individuals who are geographically dispersed or in rural areas*: A number of the programs, resources, and offices available for job placement assistance and related employment services are located on installations, in demobilization locations, or in highly populated urban centers. Several of the stakeholders we spoke with suggested that individuals who are not located near these sites (particularly individuals in rural areas) might face gaps in services because the time and effort required to access services is too great, or because they are simply not aware that the services are available. It is important for job placement assistance and related employment programs and resources to be made as accessible as possible to these populations.[7]

---

[5] Discussion with federal program manager, March 31, 2015.

[6] See Werber, Schaefer, et al., 2013.

[7] Public Law 112-260, Dignified Burial and Other Veterans' Benefits Improvement Act of 2012, January 10, 2013, authorized the Secretary of Labor to provide a pilot program that generally follows the content of TAP under 10 U.S.C. 1144 (2012) to veterans and their spouses at locations other than military installations. This two-year Off-Base Transition Training pilot was conducted in three states (Georgia, Washington, and West Virginia) to assess the feasibility and advisability of providing such a program to eligible individuals at locations other than military installations. The final report of this pilot project is scheduled for fall of 2015.

Summary

As documented in Chapter Two, there is a wide range of job placement assistance and related employment programs, resources, and offices available to reserve component members. These programs provide a range of services and are targeted to different populations, resulting in a complex network of programs that might be challenging for service members to access and challenging for service providers to utilize while providing referrals to reserve component members. There is a need to provide reserve component members with a clear way to navigate this network of programs and resources. Without a comprehensive needs assessment and evaluations of program effectiveness, we cannot fully identify gaps, but we generally found that no other new programs were needed. However, we caution that gaps could appear as programs are streamlined and consolidated. We also suggest that continued access to high-touch programs, strong connections to employers, and data-collection and case-management capabilities are essential to retain.

We also found several potential areas of overlap. We found that there were a large number of high-touch programs providing basic employment services, and while many of these programs target different populations and offer unique approaches to service provision, these programs could potentially represent overlap. In addition, there are many tools and resources that are slight variations on very similar products, and these resources could potentially be streamlined to cut costs and ensure consistency in information. Finally, while outside of the scope of the study, overlaps with state, local, and nongovernmental programs must also be taken into account when planning and implementing federal employment programs for reserve component members.

With regard to populations served, we did not find any gaps in the current portfolio of federal employment programs. Reserve component members are broadly served by programs across several agencies and are explicitly targeted by seven programs under DoD. There are some populations that could potentially be better served or that might be at-risk if programs are streamlined and consolidated without explicit consideration of retaining services to the population. These include reserve component personnel without veteran status, younger service members, individuals who are underemployed, and individuals who are geographically dispersed or in rural areas. If programs are streamlined or consolidated, special attention should be paid to these populations to ensure that they do not fall through the cracks.

CHAPTER FOUR

Recommendations for Improving Federal Job Placement Assistance for Reserve Component Members

Given our findings, outlined in Chapters Two and Three, we offer the following recommendations for improving federal job placement assistance for reserve component members:

- Identify the employment needs of reserve component members.
- Evaluate the effectiveness of existing programs in meeting the employment needs of reserve component members.
- Assess the costs and benefits of streamlining existing programs and resources.
- Consider the pros and cons of moving primary responsibility for reserve component employment assistance from DoD to other federal departments.
- Make existing programs more applicable to the reserve components and increase awareness of them
- Improve coordination and information-sharing across programs.

We discuss each of these recommendations and their implications (including cost and feasibility) below. These recommendations also provide a road map for improving federal employment assistance to reserve component members.

A Road Map for Improving Federal Employment Assistance to Reserve Component Members

Identify the Employment Needs of Reserve Component Members

While we are aware that some individual programs have tried to assess the employment needs of the general populations that they serve, our literature review and our discussions with stakeholders indicate that the employment needs of reserve component members have not been systematically identified or updated regularly. Without such an identification of the employment needs of reserve component members, it is difficult to determine whether (1) program activities are effectively serving the employment needs of reserve component members, and (2) reserve component members might have unique needs that cannot be met by more-general employment programs. Ultimately, program activities should be driven by the needs that are identified in such an assessment, and those needs should be the basis on which any future program evaluations or changes in programs are grounded.

There are many potential ways that DoD could proceed with such a needs assessment. For example, DoD could conduct surveys, interviews, or focus groups with reserve component members to help identify their employment needs, their awareness and use of different types of programs, and their opinions about these programs' effectiveness. DoD could structure the samples of these data-collection efforts to ensure the inclusion of particularly vulnerable reserve component members (e.g., junior enlisted reserve component members, reserve component members who do not live near an installation). Previous RAND research has also found that some state employment programs for reserve component members are particularly agile in meeting the changing needs of their target population because they are quickly able to identify emerging needs from the data that they already collect and track (e.g., hotline calls, utilization rates of particular services).[1]

It is also critical to recognize that pursuing cost reductions or consolidating or streamlining programs before assessing the needs of reserve component members can be counterproductive. Performing a comprehensive needs assessment will itself cost money and take time. However, the important output of the data-collection effort will be a comprehensive, up-to-date assessment of the employment needs of reserve component members, and benefits will derive from avoiding duplicate, and possibly ineffective, programs in the future.

Evaluate the Effectiveness of Existing Programs in Meeting the Employment Needs of Reserve Component Members

Once the current employment needs are identified, we recommend that the federal government conduct in-depth program evaluations of the effectiveness of major federal programs in meeting the employment needs of reserve component members (e.g., H2H, TAP, the VEC). Such formal program evaluations should include identifying program goals and outcomes (including the impact that the programs have on increasing employment opportunities and results for reserve component members, as well as their effectiveness in meeting the reserve component needs identified through our first recommendation). The evaluations would also facilitate the identification of strengths and weaknesses within federal programs, as well as gaps and overlaps across federal programs.

One aspect of many high-touch programs is the ability to track data, including individual contact information, progress on various aspects of the job search, and eventual employment outcomes. Several government programs do not collect data that differentiate veteran versus nonveteran active duty, reserve, and National Guard status. Collecting and reporting program data to more specifically differentiate veterans from nonveterans, and to distinguish active duty from reserve and National Guard members, should be a priority for high-touch programs. Case-management data can also help programs efficiently and effectively deliver high-touch services to service members, allowing counselors to follow up with individuals and tailor services to their needs.

Any program evaluation should include a specific task devoted to calculating the cost of the programs. Comparing costs of different programs will allow the federal government to evaluate existing and future programs on a consistent basis. As with our first recommendation, conducting formal program evaluations should not be predicated on searching for cost reductions. Performing the evaluations themselves will cost money and take time. However, these program evaluations will be critical to identifying how effective federal programs are in meeting the employment needs of reserve component members.

---

[1]   See Werber, Schaefer, et al., 2013.

Assess the Costs and Benefits of Streamlining Existing Programs and Resources

Based on the information collected in our first two recommendations, DoD and the rest of the federal government should then assess the costs and benefits of streamlining existing federal programs and resources. Both the needs assessment and the program evaluations will provide information that will help to determine whether streamlining programs and resources should be carried out and, if so, which programs and resources should be streamlined.

Streamlining could have many benefits. It could potentially reduce costs by eliminating redundant services, tools, and resources. For instance, programs that have similar (or identical) technical services could potentially reduce operating and maintenance costs if they were centralized and managed jointly. Streamlining could also potentially alleviate confusion among service members because of too much or conflicting information.

However, before streamlining any overlapping services, it is critical to first identify whether those overlaps exist for a reason (e.g., to meet different needs or serve different populations). Seemingly redundant services might not be a problem if they are addressing unique needs or serving different populations. It is also important to recognize that while streamlining programs and resources could reduce salary costs, it could also limit service delivery to some populations. Therefore, it is essential to ensure that the needs of reserve component members continue to be met during any consolidation of programs.

Consider the Pros and Cons of Moving Primary Responsibility for Reserve Component Employment Assistance from DoD to Other Federal Departments

Given that DoD's primary mission is warfighting, some might argue that it is appropriate to move primary responsibility for reserve component employment assistance away from DoD to other federal departments, such as DoL and the VA, whose primary missions align more closely with providing employment assistance to all Americans or social services to veterans. Simply adding reserve component employment needs to programs that already serve broader populations might save costs, streamline services, and potentially make it easier for service members to navigate employment resources.

However, these benefits should be weighed carefully with the potential downsides of moving primary responsibility for reserve component employment assistance away from DoD to other federal departments. For instance, DoD has the best access to its reserve component population and can potentially collect vital data on the needs of this population more easily than other federal agencies can. In addition, DoD has already built a sophisticated infrastructure (through its H2H program) to address the employment needs of reserve component members.

The key issue is to identify whether there are unique reserve component challenges that warrant a parallel set of employment services specifically for reserve component members. Given the multiple constraints placed on this study, we are not able to provide a recommendation as to whether DoD should retain primary responsibility for reserve component employment assistance, or whether that responsibility should be moved to another federal agency. However, such a decision could and should be made after a needs assessment has been conducted, the major federal employment programs that serve reserve component members have been formally evaluated, and the costs and benefits of streamlining existing programs have been assessed.

At the very least, we recommend that if primary responsibility for the administration of reserve component employment assistance is ultimately moved from DoD to another federal department, a DoD position should be established to monitor employment services for reserve component members in non-DoD federal departments. This should not simply be a liaison position but, rather, a formal full-time position with the associated responsibilities, authorities, funding, and accountability to ensure that the employment needs of reserve component members are being met within the context of a broader employment program in a non-DoD agency. This could be a feasible option for delegating the administration of employment programs to non-DoD agencies, while maintaining DoD oversight over their effectiveness in meeting the needs of reserve component members.

In addition, we recommend that particular reserve component populations (e.g., reserve component personnel without veteran status, young reserve component members, underemployed individuals, and individuals who are geographically dispersed or in rural areas) be carefully assisted to ensure that they do not fall through the cracks during any potential program transitions. The identification of the needs of these populations would be part of the needs assessment discussed in our first recommendation. Similarly, we recommend that some key employment services (e.g., high-touch programs, strong connections to employers, and data-collection and case-management capabilities) be retained in any program transition to ensure that the needs of reserve component members continue to be met.

Make Existing Programs More Applicable to the Reserve Components and Increase Awareness of Them

One of the most feasible, short-term steps that DoD (and the federal government as a whole) could take to improve employment assistance for reserve component members is to make the most of programs that already exist and to increase awareness of those programs among reserve component members. This could prevent the proliferation of additional, potentially redundant, programs. Most important, the federal government should consider making existing assistance (such as TAPs and AJCs) more applicable to reserve component members. This could increase program impacts, reduce costs, and provide consistent information across the components.

As our study indicates, there are a large number of federal job placement and employment programs, resources, and offices available to reserve component members. Simply increasing awareness of these existing programs could potentially have a positive impact on reserve component members' employment outcomes. As indicated in Chapter Three, this might involve providing reserve component members with additional guidance or more-streamlined information about the available federal programs and resources. However, increasing the utilization of existing programs might stem the tide of demand for the establishment of potentially redundant new programs and services. In addition, the cost of any program is spread over the number of people who utilize it. While some variable costs might accrue for each additional user, the amount of fixed costs (both up-front and recurring) per user automatically decreases as more people take advantage of the program.

Improve Coordination and Information-Sharing Across Programs

Finally, we recommend that the federal government improve coordination and information-sharing across employment programs (especially about activities). One of the major findings from our informational discussions is that some programs appear to have very little visibility on other programs' activities. Thus, it is not a surprise that there are some overlaps across programs. Coordination could be increased quickly through very feasible, low-cost measures, such as periodic meetings among program staff, where they can share information about issues, including the types of employment needs that their programs are seeing among reserve component members, what their programs are currently doing to address those needs, and any planned changes in their programs' activities.

While increased information-sharing can potentially introduce complex legal, policy, and data-management issues (e.g., with respect to personally identifiable information), sharing information about program activities, at the very least, could increase program impacts, save resources, and prevent overlaps. Additional forms of information-sharing will require the necessary changes to legal authorities, policy, and data-management systems.

Programs can make better decisions when they have better information and can avoid costly replication of existing assets or services. Sharing relevant information can also reduce personnel costs. The key word, however, is *relevant*. Not all data should be disseminated among multiple programs, and indeed cannot be made open because of legal or policy concerns. A blanket declaration that data should be shared might not help reduce costs; assessing information needs, as well as collection, storage, and sharing protocols, should be part of the overall program evaluations that we recommend.

Things to Consider Before Making Changes

In light of our study findings, we also identify several items that DoD and the rest of the federal government should consider before making changes to the current federal portfolio of employment assistance to reserve component members. These include

- assess the impact of potential changes to employment programs or activities
- ensure that new programs will not overlap with existing programs and activities
- recognize that cost is only one aspect of effectiveness
- plan up front to facilitate cost measurement and comparison across programs.

These measures are vital steps in facilitating the increased effectiveness of federal employment programs that assist reserve component members. The measures will also help to ensure that any changes to those programs' activities will be guided by the needs of reserve component members and that those changes will decrease duplication and instead fill potential gaps in services. We discuss each of these in more detail below.

Assess the Impacts of Potential Changes

Before any additional changes are made to federal employment programs or funding, it is critical to first assess the impacts of those potential changes. Most important, before any changes are made, the federal government as a whole should ensure that the unique services they provide to reserve component members will not be lost when the changes are made. It should also ensure that particular populations within the reserve components (e.g., reserve component personnel without veteran status, young reserve component members, underemployed individuals, and individuals who are geographically dispersed or who live in rural areas) do not fall through the cracks when changes are made.

Ensure That New Programs Will Not Overlap with Existing Programs and Activities

Our discussions with stakeholders indicate that, in some cases, they are not aware of what other programs are doing and how their own program activities might overlap with others. Consequently, any program that is contemplating adding or changing services should first check to see whether other programs already have duplicate resources or resources that can be built on, rather than starting from scratch. This can also reduce overall costs to the federal government, since it will avoid paying for the same thing more than once, and will save programs operating time and costs.

Recognize That Cost Is Only One Aspect of Effectiveness

While there will inevitably be pressure to reduce costs within and across federal employment programs, it is important to recognize that evaluating cost effectiveness is not always a simple comparison, because the underlying causes of those costs vary in important ways. For example, research indicates that the time people spend in training and placement programs is a function of their individual preparedness for the workforce.[2] Those who are not as prepared tend to stay longer, making them appear more expensive to the program. However, it is precisely these people who are receiving the most value from the program, since those who leave earlier were likely to be better prepared. The dynamic nature of effectiveness, as well as costs and benefits, leads directly to our next suggestion.

Plan Up Front to Facilitate Data Collection and Comparison Across Programs

Program effectiveness should be tracked over time. The needs assessment and program evaluations recommended in this study will produce a view of the current landscape, but they are not dynamic models. Part of the development of current programs should include creating guidelines for the kinds of data that should be routinely collected and how those data should be used. Without identifying common denominators or outcomes, comparing across programs is impossible, making any attempt to assess effectiveness fruitless.

Such data will also facilitate cost comparisons across programs and resources. Collecting costs on a regular basis can impose new requirements on a program but are worth the extra effort. Once the common denominator has been established, a uniform list of data to track should be provided to the programs. Those data should include the major items in the cost element structure: These are the direct costs that are incurred by the program. The data should also include measures by which effectiveness is assessed. This might include the number of people who visit an office (if one exists), the number of registered users on a website (if one exists), the number of requests for information made by service members and relations (depending on the program), the number of service members acquiring a job, and any other relevant pieces of information. Collecting more data than just the single metric for the cost ratio allows for flexibility during review. The metrics can be used by the program for ongoing program improvements and also by the federal government for oversight.

Programs designed to assist job seekers strive for multiple positive effects. Collecting cost information provides information only for a specific outcome. Evaluating a program's success, then, involves more than just tallying costs. But when costs are factored into the evaluation,

---

2   Robert J. LaLonde, "Employment and Training Programs," in Robert A. Moffitt, ed., *Means-Tested Transfer Programs in the United States*, Chicago: University of Chicago Press, 2003.

they should provide evaluators confidence about what the costs reflect and allow for principled analysis. The items noted above will help accomplish this task.

Final Thoughts

Given recent changes in the U.S. economy, the operational tempo of deployments, and the addition of new federal employment programs, this is an opportune time for the federal government to reevaluate ways in which it can improve federal job placement and related employment services for reserve component members. While we did not identify major gaps in services provided to reserve component members, we caution that our findings are based on limited data and that the context within which employment challenges evolve is very dynamic. Therefore, additional data collection and analysis are necessary before any major changes are made to the current constellation of federal employment programs in an effort to improve those programs and resources for reserve component members. The recommendations in this study provide a road map for ways to reevaluate and improve the current portfolio of federal employment programs and resources available to reserve component members.

Summary Table of Federal Job Assistance Programs, Resources, and Offices

Agency	Program or Resource	Population Served	Services Provided	Intensity of Services
DoD	Beyond the Yellow Ribbon (BYR) programs	Members of the reserve components in states receiving program funding	Varies by state program; provides funding to employment programs that offer one-on-one assistance with job search, job posting, résumé and interview prep, and outreach to and follow-up with employers	Varies by program
	DoD Hiring Heroes	Service members, veterans, spouses, and primary caregivers; unclear if members of the reserve components are eligible	One-on-one assistance by toll-free hotline, chat sessions, or email to receive guidance on DoD employment and help with vacancies and application processes; links to tools, information, and programs	Varies
	Education and Employment Initiative	All service members with service-related injuries	One-on-one education and career counseling; assessment of work competencies	High touch
	Hero2Hired (H2H)	Members of the reserve components	One-on-one assistance with job search, assistance with job postings, résumé and interview prep; outreach to and follow-up with employers	High touch
	Military OneSource	All service members and their families	One-on-one assistance with all benefit-related issues including education and career support; information for employers and service providers	Low touch
	Operation Warfighter (OWF)	All service members with service-related injuries	Training; apprenticeships; internships; one-on-one assistance with matching to positions; and résumé preparation	High touch
	SkillBridge	Service members returning from deployment and separating; more than 180 days of active duty	Training; apprenticeships; internships	High touch
	Troops to Teachers (TTT)	All current and former service members with honorable service	Counseling on certification requirements and employment leads; links to state resources; financial assistance (for a limited number of participants)	Low touch
	Yellow Ribbon Reintegration Program (YRRP)	Members of the reserve components, spouses	Events that provide information about a range of benefits and access to employers and service providers; courses on employment-related issues	Varies

Agency	Program or Resource	Population Served	Services Provided	Intensity of Services
DoD and Air Force	Airman and Family Readiness	Active duty service members, reserve component members, civilians, retirees, and their families	One-on-one assistance with job search; access to job postings; résumé and interview prep; resources on employment tools and programs	High touch
DoD and Army	Always a Soldier (AAS)	All service members with service-related injuries		Low touch
DoD and Army	Career Skills Program (CSP)	Veterans separating from the military within 180 days; at least one year of active duty or two years of reserve service	Apprenticeships, on-the-job training, job shadowing, employment skills training, internships	High touch
DoD, Army, Navy, Air Force, and Marine Corps	Credentialing Opportunities Online (COOL)	Service members and service providers	A tool (search engine) that provides information on credentialing requirements for occupations related to a civilian occupation; identifies licenses and certifications related to specific military specialties; provides funding for some credentialing activities	Information and tools (some have low-touch services in the form of financial support)
DoD and Army	Employment Readiness Program (ERP)	Focuses on military spouses, but also available for all service members, retirees, and DoD civilians	One-on-one assistance with job search, access to job postings, résumé and interview prep; courses on employment-related topic; job fairs	High touch
DoD and Navy	Fleet and Family Support Centers	All service members in the Navy and their families	One-on-one employment assistance; career courses; TAP workshops; counseling for health, sexual assault; relocation assistance	High touch
DoD and Army National Guard	Guard Apprenticeship Program Initiative (GAPI)	Members of the Army National Guard and Army Reserve	Apprenticeships and on-the-job training	High touch
DoD and National Guard	Job Connection Education Program (JCEP)	Members of the reserve components and their spouses	One-on-one assistance with job search, access to job postings, résumé and interview prep; establishing relationships with employers; education and training assistance	High touch
DoD and National Guard	National Guard Employment Network (NGEN)	Primarily guard members and their families, but also reservists and veterans	One-on-one assistance with job search, access to job postings, résumé and interview prep; database for case management of service members and employers; access to information and tools	High touch
DoD and Marine Corps	Marine and Family Program's Career Services	All service members, retirees, veterans, DoD civilians, their families	One-on-one assistance with job search, job posting, résumé and interview prep; access to resources; workshops and employer events	High touch
DoD and Army	Partnership for Youth Success (PaYS)	New active and reserve Army enlistees	A guaranteed interview and possible job with an employer after separation	Low touch
DoD and Army Reserve	Private Public Partnership (P3) program	Active component and reserve component soldiers	One-on-one assistance with job search, access to job postings, résumé and interview prep; establishing relationships with employers	High touch

Agency	Program or Resource	Population Served	Services Provided	Intensity of Services
DoL	American Job Centers (AJCs)	All Americans (veterans given priority for services through Gold Card)	For veterans, one-on-one assistance with job search, access to job postings, résumé and interview prep; access to information and tools; case management; toll-free hotline for assistance with resources	Low touch (high touch for veterans)
	Career OneStop	All Americans	Access to information and tools; toll-free hotline for assistance with resources	Low touch
	O*NET and My Next Move—For Veterans	Veterans, all service members, the public	Search engines that provide information about occupations; access to tools and assessments for career information; information about job preparation (e.g., résumés)	Information and tools
	Registered Apprenticeship	All Americans	Apprenticeships	High touch
	Veterans' Employment and Training Service (VETS)	Veterans, all service members, families	Oversees AJCs, TAP program, My Next Move, Women Who Serve; access to fact sheets, employment programs, legal information	Low touch
VA	Compensated Work Therapy (CWT)	Veterans with disabilities	One-on-one treatment and case management; simulated work training; vocational assessment; assistance with job access; rehabilitation in residential settings	High touch
	Education and Career Counseling and CareerScope	Transitioning active duty service members, veterans, their dependents	Career assessment; educational and career counseling on occupations and education programs	Uncertain; additional information needed
	Office of Small and Disadvantaged Business Utilization (OSDBU)	Veterans, family members	Online information and resources; access to counselors to assist with business counseling and verification	Low touch; information and tools
	VA for Vets	Veterans and all service members interested in working for the VA	Resources and information on employment at the VA; call line with access to assistance on VA employment; information for human resources professionals	Information and tools
	Veterans Employment Center (VEC)	Veterans, all service members, families, authorized caregivers	informational resources and tools (résumé builder, job-posting board, skill translator); links to VA programs; information for employers	Information and tools
	Veteran's Opportunity to Work (VOW)	Active duty and veterans with disabilities	One-on-one vocational rehabilitation services; funding for employers; overseeing TAP; employment information and links to other programs	Low touch for all veterans; high touch for qualifying service-disabled veterans

Agency	Program or Resource	Population Served	Services Provided	Intensity of Services
VA	VetSuccess on Campus (VSOC)	Transitioning active duty service members, student veterans who are receiving education benefits	Community and on-campus outreach, educational and vocational assessments and counseling, and referrals to other programs and services	Uncertain; additional information needed
	Vocational Rehabilitation and Employment (VR&E)	Transitioning active duty service members, veterans with disabilities	One-on-one rehabilitation services including assistance with job search, access to job postings, resume and interview prep; apprenticeship and internships; online resources and tools	High touch
	Workstudy Program	Service members, veterans, and families receiving education benefits and enrolled at least 75 percent of the time	Placement into part-time employment during educational studies	Low touch
Office of Personnel Management	Feds Hire Vets	Veterans, transitioning service members, families	Information and resources for individuals seeking federal employment; training modules; links to other programs and tools; information for HR professionals	Information and tools
Small Business Administration	Veterans Business Outreach Program (VBOP)	Veterans, members of the reserve components, families	One-on-one services for entrepreneurs, including business training, counseling and mentoring, and referrals	Some high-touch services and information and tools
DoD, DoL, and VA	National Resource Directory	Veterans, all service members, families	Search engine to identify programs and service providers in a range of benefit areas	Information and tools
	Transition Assistance Program (TAP)	Transitioning active duty service members and reserve component members demobilizing after at least 180 days of active service	Events that provide information on a range of benefits and access to employers and service providers; 5-day course on employment-related issues; employment, training, and entrepreneurship tracks	Low touch
All federal agencies	Veterans Employment Program Offices (VEPOs)	Veterans, families	Employment-related information, resources, and program links; information sessions; some offices provide access to one-on-one support including assistance with job search, access to job postings, resume and interview prep	Information and tools (some have low-touch services)

Literature Review Data-Abstraction Form

NAME (WEBSITE):

OVERVIEW:

- Which federal department/subcomponent of the department administers the program
- How the program helps guard and reserve members
- Program managers, titles, contact information

TARGET POPULATION/ELIGIBILITY:

- Who is eligible and what the eligibility requirements are:
  - How long someone can be eligible?
  - How soon can someone start using the program (e.g., do they have to be near the end of their tenure as guard/reserve)?
  - What does someone have to do to participate in the program (e.g., enrollment)?

ACTIVITIES:

- How the program impacts an increase in employment activities
- What the program does—what types of activities the program provides
- Events
- Featured jobs

COMPARISON TO OTHER PROGRAMS:

- Limitations: What the program does not do
- Overlaps with other programs
- Gaps between programs

Informational Discussion Protocol

Overview/Introduction

1. To start, please tell us your job title and main responsibilities at [PROGRAM NAME].
2. When did [PROGRAM NAME] start?
   a. Probe: Did it operate under another name before, or as part of a larger program? If yes, please tell me about how [PROGRAM NAME] evolved to its present state.
3. What are [PROGRAM NAME'S] goals and objectives?

Target Population

4. We'd like to learn more about [PROGRAM NAME'S] target population. Would you please tell me who is part of the target population, and how large that population is?
   a. Prompt if needed: An estimate of the population size is sufficient.
5. Are guard and reserve members the sole target audience, or do other types of people participate as well?
6. How many [target population] does [PROGRAM NAME] serve? [If not obvious:] Is that measured per year, month, week, day?
7. How broadly is [PROGRAM NAME] currently being implemented?
   a. Prompt if needed: For example, how many locations does the program have? Are all eligible? Who is excluded?
   b. Prompt if needed: When does someone become eligible for the program? How long are they eligible? What (if anything) do they need to do in order to participate?

Program Activities

8. Our next set of questions is intended to help us understand what [PROGRAM NAME] does to achieve the goals you mentioned earlier. First, what activities or services constitute [PROGRAM NAME]?
9. [If not answered above:] What types of problems does [PROGRAM NAME] address most frequently in these activities and services?
10. How does your target audience learn about these activities and services? What kinds of outreach does [PROGRAM NAME] do?
    a. Do you attend any military-affiliated events to advertise [PROGRAM NAME'S] services?

11. Have any of [PROGRAM NAME'S] activities or strategies been particularly success-ful? Why?

Comparison to Other Employment Programs

12. What other programs directly or indirectly provide job assistance to guard and reserve members?
    a. Prompt: Ask for a list of names, target populations, objectives.
    b. Prompt: Are there programs that don't specifically target guard and reserve mem-bers, but for which they are eligible?

13. Now we want to ask how *your* program compares to these other programs designed to provide job placement and other related services to guard and reserve members (or for which guard and reserve members are eligible) (see below)
    a. How is your program *unique or different* from the other programs you have listed?
       ◦ Goals and objectives?
       ◦ Activities and services?
       ◦ Target population?
       ◦ Other?
    b. How is your program *similar to* the other programs you have listed?
       ◦ Goals and objectives?
       ◦ Activities and services?
       ◦ Target population?
       ◦ Other?

14. Are there *overlaps* between some programs designed to provide job placement and other related services to guard and reserve members? If so, what are they? How extensive or important are these overlaps?

15. Are there *gaps* across programs designed to provide job placement and other related ser-vices to guard and reserve members? If so, what are they? How extensive or important are these gaps?
    a. Are there any important employment services that are not being provided to guard and reserve members as a result of these gaps?

16. How could programs that provide employment services for guard and reserve members improve coordination across programs?

17. What could be done to improve federal employment services for guard and reserve members?
    a. Prompt: For example, change laws or policies that reduce the effectiveness of pro-grams?

Future of the Program

18. Thank you again for your time and all the information you provided today. In closing, we'd like to discuss briefly the future of [PROGRAM NAME]. We know that [PRO-GRAM NAME] started [year of inception]. Do you plan to make any changes in the program for purposes of sustainment?

19. What plans, if any, does [PROGRAM NAME] have to grow? This could include offering more activities and services, adding locations, and/or reaching a larger target audience.
20. In closing, is there anything else you would like to tell me about [PROGRAM NAME] that we haven't discussed?
21. Is there anything else you would like to tell me about job assistance programs for guard and reserve members (in general) that we haven't discussed?

Program Data Sources

Airman and Family Readiness program website,
http://www.afrc.af.mil/AboutUs/AirmanFamily.aspx

Always a Soldier program web page,
http://www.amc.army.mil/amc/alwaysasoldier.html

American Job Centers program website,
http://jobcenter.usa.gov/

Career OneStop program website,
http://www.careeronestop.org/reemployment/veterans/

CareerScope web page,
http://www.benefits.va.gov/gibill/careerscope.asp

Career Skills Program web page,
https://www.hrc.army.mil/TAGD/Career%20Skills%20Program%20-%20CSP

Credentialing Opportunities On-Line program website,
https://www.cool.army.mil/about.htm

Education and Employment Initiative program web page,
http://warriorcare.dodlive.mil/wounded-warrior-resources/e2i/

Employer Support of Guard and Reserve program website,
http://www.esgr.mil/

Employment Readiness Program web page,
http://www.myarmyonesource.com/FamilyProgramsandServices/FamilyPrograms/EmploymentReadiness/
default.aspx

Feds Hire Vets website, http://www.fedshirevets.gov/

Fleet and Family Support Centers web page,
http://www.cnic.navy.mil/ffr/family_readiness/fleet_and_family_support_program.html

Gold Card Program web page,
http://www.dol.gov/vets/goldcard.html

Guard Apprenticeship Program Initiative program web page,
https://g1arng.army.pentagon.mil/Programs/GAPI/Pages/default.aspx

Hero2Hired program website,
http://h2h.jobs/

Hiring Heroes program web page,
http://godefense.cpms.osd.mil/veterans/hiringheroes.aspx

Job Connection Education Program website,
http://jcep.info/

Marine and Family Program's Career Services web page,
http://www.mccscp.com/career-services

Military OneSource program website,
https://www.militaryonesourceeap.org/achievesolutions/en/militaryonesource/Home.do

My Next Move for Veterans program website,
http://www.mynextmove.org/vets/

National Guard Employment Network web page,
http://www.nationalguard.com/employment-network

National Resource Directory web page,
https://www.ebenefits.va.gov/ebenefits/nrd

O*Net program website,
https://www.onetonline.org/

Operation Warfighter program web page,
http://warriorcare.dodlive.mil/wounded-warrior-resources/operation-warfighter/

PaYS program website,
https://www.armypays.com/INDEX.html

Private Public Partnership program web page,
http://www.usar.army.mil/resources/Pages/Employer-Partnership-opportunities-and-Information.aspx

SkillBridge program website,
http://www.dodskillbridge.com/

Troops to Teachers program website,
http://troopstoteachers.net/

U.S. Department of Defense Transition Assistance Program website,
https://www.dodtap.mil/

U.S. Department of Labor Registered Apprenticeship program web page,
http://www.doleta.gov/OA/veterans.cfm

U.S. Department of Labor Transition Assistance Program web page,
http://www.dol.gov/vets/programs/tap/

U.S. Department of Veterans Affairs, Office of Small and Disadvantaged Business Utilization, web page,
http://www.va.gov/osdbu/

U.S. Department of Veterans Affairs Compensated Work Therapy program web page,
http://www.va.gov/health/cwt/

U.S. Department of Veterans Affairs Transition Assistance Program web page,
http://www.benefits.va.gov/VOW/tap.asp

U.S. Department of Veterans Affairs Workstudy program web page,
http://www.benefits.va.gov/gibill/workstudy.asp

VA for Vets program website,
http://vaforvets.va.gov/

Veterans Business Outreach Program web page,
https://www.sba.gov/offices/headquarters/ovbd/resources/362341

Veterans Employment and Training Service agency website,
http://www.dol.gov/vets/

Veterans Employment Center program web page,
https://www.ebenefits.va.gov/ebenefits/jobs

Veterans Opportunity to Word program web page,
http://www.benefits.va.gov/VOW/

VetSuccess on Campus program web page,
http://www.benefits.va.gov/vocrehab/vsoc.asp

Vocational Rehabilitation and Employment program web page, http://www.benefits.va.gov/VOCREHAB/index.asp

Werber, Laura, Jennie W. Wenger, Agnes Gereben Schaefer, Lindsay Daugherty, and Mollie Rudnick, *An Assessment of Fiscal Year 2013 Beyond Yellow Ribbon Programs*, Santa Monica, Calif.: RAND Corporation, RR-965-OSD, 2015.

Yellow Ribbon Reintegration Program web page, https://www.jointservicessupport.org/yrrp

Abbreviations

AJC	American Job Center
CWT	Compensated Work Therapy
DoD	Department of Defense
DoL	Department of Labor
EC	employment coordinator
ERP	Employment Readiness Program
ESGR	Employer Support of the Guard and Reserve
FFSC	Fleet and Family Support Center
GAPI	Guard Apprenticeship Program Initiative
H2H	Hero2Hired
JCEP	Job Connection Education Program
JVSG	Jobs for Veterans State Grants Program
NDAA	National Defense Authorization Act
NGEN	National Guard Employment Network
PaYS	Partnership for Youth Success
RCS	report control symbol
TAP	Transition Assistance Program
TTT	Troops to Teachers
USERRA	Uniformed Services Employment and Reemployment Rights Act
VA	Department of Veterans Affairs
VEC	Veterans Employment Center
VEPO	Veterans Employment Programs Office
VETS	Veterans' Education and Training Service

VR&E   Vocational Rehabilitation and Employment

YRRP   Yellow Ribbon Reintegration Program

References

Acosta, Joie D., Gabriella C. Gonzalez, Emily M. Gillen, Jeffrey Garnett, Carrie M. Farmer, and Robin M. Weinick, *The Development and Application of the RAND Program Classification Tool: The RAND Toolkit, Volume 1*, Santa Monica, Calif.: RAND Corporation, RR-487/1-OSD, 2014. As of August 11, 2015: http://www.rand.org/pubs/research_reports/RR487z1

"Air Force Airman and Family Readiness," web page, Air Force Reserve Command, n.d. As of August 11, 2015: http://www.afrc.af.mil/AboutUs/AirmanFamily.aspx

American Job Centers, website, n.d. As of August 11, 2015: http://jobcenter.usa.gov/

"Army Always a Soldier," web page, U.S. Army Materiel Command, n.d. As of August 11, 2015: http://www.amc.army.mil/amc/alwaysasoldier.html

"Army Materiel Command's Always a Soldier Program," web page, Career Center for Wounded Warriors and Disabled Veterans, Military.com, n.d. As of August 11, 2015: http://www.military.com/hero/0,,WVC_AlwaysASoldier,00.html

Army Partnership for Youth Success, website, n.d. As of August 11, 2015: https://www.armypays.com/INDEX.html

Career OneStop, website, n.d. As of August 11, 2015: http://www.careeronestop.org/reemployment/veterans/

"CareerScope," web page, U.S. Department of Veterans Affairs, n.d. As of August 11, 2015: http://www.benefits.va.gov/gibill/careerscope.asp

"Career Services," web page, Marine and Family Programs, Marine Corps Community Services Camp Pendleton, n.d. As of August 11, 2015: http://www.mccscp.com/career-services

"Career Skills Program—CSP," web page, United States Army Human Resources Command, n.d. As of August 11, 2015: https://www.hrc.army.mil/TAGD/Career%20Skills%20Program%20-%20CSP

"Compensated Work Therapy," web page, U.S. Department of Veterans Affairs, n.d. As of August 11, 2015: http://www.va.gov/health/cwt/

Credentialing Opportunities On-Line, website, n.d. As of August 11, 2015: https://www.cool.army.mil/about.htm

DoD—*See* U.S. Department of Defense.

"DoD Hiring Heroes Program," web page, U.S. Department of Defense, n.d. As of August 11, 2015: http://godefense.cpms.osd.mil/veterans/hiringheroes.aspx

DoD SkillBridge, website, n.d. As of August 11, 2015: http://www.dodskillbridge.com

Dreisbach, Tom, and Rachel Martin, "National Guard Members' Next Battle: The Job Hunt," *Weekend Edition*, NPR.org, April 29, 2012. As of August 11, 2015: http://www.npr.org/2012/04/29/151619099/national-guard-members-next-battle-the-job-hunt

"Education and Employment Initiative," web page, *Warrior Care Blog*, DoD Office of Warrior Care Policy, n.d. As of August 11, 2015:
http://warriorcare.dodlive.mil/wounded-warrior-resources/e2i/

Employer Support of the Guard and Reserve, website, n.d. As of August 11, 2015:
http://www.esgr.mil/

"Employment Readiness Program," web page, Army OneSource, n.d. As of August 11, 2015:
http://www.myarmyonesource.com/FamilyProgramsandServices/FamilyPrograms/EmploymentReadiness/default.aspx

Executive Order 13518, *Employment of Veterans in the Federal Government*, Washington, D.C.: The White House, Office of the Press Secretary, November 9, 2009.

Feds Hire Vets, website, n.d. As of August 11, 2015:
http://www.fedshirevets.gov/

"Fleet and Family Support Centers," web page, Commander, Navy Installations Command, n.d. As of August 11, 2015:
http://www.cnic.navy.mil/ffr/family_readiness/fleet_and_family_support_program.html

Gates, Susan, Geoffrey McGovern, Ivan Waggoner, John D. Winkler, Ashley Pierson, Lauren Andrews, and Peter Buryk, *Supporting Employers in the Reserve Operational Forces Era*, Santa Monica, Calif.: RAND Corporation, RR-152-OSD, 2013. As of August 11, 2015:
http://www.rand.org/pubs/research_reports/RR152

"Guard Apprenticeship Program Initiative (GAPI)," web page, July 8, 2011. As of August 11, 2015:
https://g1arng.army.pentagon.mil/Programs/GAPI/Pages/default.aspx

Hall, Kimberly Curry, Margaret C. Harrell, Barbara A. Bicksler, Robert Stewart, and Michael P. Fisher, *Veteran Employment: Lessons from the 100,000 Jobs Mission*, Santa Monica, Calif.: RAND Corporation, RR-836-JPMCF, 2014. As of August 11, 2015:
http://www.rand.org/pubs/research_reports/RR836.html

Hero2Hired, website, n.d. As of August 11, 2015:
http://h2h.jobs/

"Job Seekers," job bank, Veterans Employment Center, eBenefits, n.d. As of August 11, 2015:
https://www.ebenefits.va.gov/ebenefits/jobs

LaLonde, Robert J., "Employment and Training Programs," in Robert A. Moffitt, ed., *Means-Tested Transfer Programs in the United States*, Chicago: University of Chicago Press, 2003, pp. 517–586.

Military OneSource, website, n.d. As of August 11, 2015:
https://www.militaryonesource.mil

My Next Move for Veterans, website, n.d. As of August 11, 2015:
http://www.mynextmove.org/vets/

"National Guard Employment Network," web page, National Guard, n.d. As of August 11, 2015:
http://www.nationalguard.com/employment-network

National Guard Job Connection Education Program, website, n.d. As of August 11, 2015:
http://jcep.info/

"National Resource Directory," web page, eBenefits, Department of Veterans Affairs and the Department of Defense, n.d. As of August 11, 2015:
https://www.ebenefits.va.gov/ebenefits/nrd

"New Employment Initiatives for Veterans," web page, U.S. Department of Labor, n.d. As of August 11, 2015:
http://www.dol.gov/vets/goldcard.html

O*Net Online, website, n.d. As of August 11, 2015:
https://www.onetonline.org/

"Office of Small and Disadvantaged Business Utilization," web page, U.S. Department of Veterans Affairs, n.d. As of August 11, 2015:
http://www.va.gov/osdbu/

Office of the First Lady, White House, "Obama Administration Launches Online Veterans Employment Center: One-Stop-Shop Connects Veterans, Transitioning Service Members, and Their Spouses to Employers," April 23, 2014. As of August 12, 2015:
https://www.whitehouse.gov/the-press-office/2014/04/23/obama-administration-launches-online-veterans-employment-center-one-stop

Office of the Secretary of Defense for Reserve Affairs, *Benefits Guide*, Washington, D.C.: U.S. Department of Defense, September 2012. As of August 12, 2015:
http://ra.defense.gov/Portals/56/Documents/Benefits%20Guide%202012.pdf

"Operation Warfighter," web page, *Warrior Care Blog*, DoD Office of Warrior Care Policy, n.d. As of August 11, 2015:
http://warriorcare.dodlive.mil/wounded-warrior-resources/operation-warfighter/

Pint, Ellen M., Amy Richardson, Bryan W. Hallmark, Scott Epstein, and Albert L. Benson, *Employer Partnership Program Analysis of Alternatives*, Santa Monica, Calif.: RAND Corporation, TR-1005-A, 2012. As of August 12, 2015:
http://www.rand.org/pubs/technical_reports/TR1005

Prince, Paul D., "Out of Sight but Not Forgotten; Study Looks at Geographically Dispersed Soldiers, Families," Army.mil, October 7, 2009. As of August 12, 2015:
http://www.army.mil/article/28460/Out_of_sight_but_not_forgotten__Study_looks_at_geographically_dispersed_Soldiers__Families/

"Private Public Partnerships," web page, U.S. Army Reserve, n.d. As of August 11, 2015:
http://www.usar.army.mil/resources/Pages/Employer-Partnership-opportunities-and-Information.aspx

Public Law 112-260, Dignified Burial and Other Veterans' Benefits Improvement Act of 2012, January 10, 2013.

Public Law 113-291, Carl Levin and Howard P. "Buck" Mckeon National Defense Authorization Act for Fiscal Year 2015, Section 583, December 19, 2014.

"Registered Apprenticeship," web page, U.S. Department of Labor, n.d. As of August 11, 2015:
http://www.doleta.gov/OA/veterans.cfm

"Transition Assistance Program," web page, U.S. Department of Labor, n.d. As of August 11, 2015:
http://www.dol.gov/vets/programs/tap/

"Transition Assistance Program," web page, U.S. Department of Veterans Affairs, n.d. As of August 11, 2015:
http://www.benefits.va.gov/VOW/tap.asp

Transition Assistance Program, website, U.S. Department of Defense, n.d. As of August 11, 2015:
https://www.dodtap.mil/

Troops to Teachers, website, n.d. As of August 11, 2015:
http://troopstoteachers.net/

United States Code, Title 10, Section 1144, Employment Assistance, Job Training Assistance, and Other Transitional Services: Department of Labor, January 3, 2012.

United States Code, Title 38, Section 4211, Definitions, January 3, 2012.

U.S. Department of Defense Instruction 8910.01, *Information Collection and Reporting*, Washington, D.C.: U.S. Department of Defense, May 19, 2014. As of August 13, 2015:
http://www.dtic.mil/whs/directives/corres/pdf/891001p.pdf

U.S. Government Accountability Office, *Fragmentation, Overlap, and Duplication: An Evaluation and Management Guide*, Washington, D.C., GAO-15-49SP, 2015.

VA for Vets, website, U.S. Department of Veterans Affairs, n.d. As of August 11, 2015:
http://vaforvets.va.gov/

"Veterans Business Outreach Program," web page, U.S. Small Business Administration, n.d. As of August 11, 2015:
https://www.sba.gov/offices/headquarters/ovbd/resources/362341

Veterans Employment and Training Service, website, U.S. Department of Labor, n.d. As of August 11, 2015:
http://www.dol.gov/vets/

"Veterans Opportunity to Work," web page, U.S. Department of Veterans Affairs, n.d. As of August 11, 2015:
http://www.benefits.va.gov/VOW/

"VetSuccess on Campus," web page, U.S. Department of Veterans Affairs, n.d. As of August 11, 2015:
http://www.benefits.va.gov/vocrehab/vsoc.asp

"Vocational Rehabilitation and Employment (VR&E)," web page, U.S. Department of Veterans Affairs, n.d. As of August 11, 2015:
http://www.benefits.va.gov/VOCREHAB/index.asp

Vogel, Steve, "Returning Military Members Allege Job Discrimination—by Federal Government," *The Washington Post*, February 19, 2012.

Werber, Laura, Agnes Gereben Schaefer, Karen Chan Osilla, Elizabeth Wilke, Anny Wong, Joshua Breslau, and Karin E. Kitchens, *Support for the 21st Century Reserve Force: Insights on Facilitating Successful Reintegration for Citizen Warriors and Their Families*, Santa Monica, Calif.: RAND Corporation, RR-206-OSD, 2013. As of August 11, 2015:
http://www.rand.org/pubs/research_reports/RR206

Werber, Laura, Jennie W. Wenger, Agnes Gereben Schaefer, Lindsay Daugherty, and Mollie Rudnick, *An Assessment of Fiscal Year 2013 Beyond Yellow Ribbon Programs*, Santa Monica, Calif.: RAND Corporation, RR-965-OSD, 2015. As of August 11, 2015:
http://www.rand.org/pubs/research_reports/RR965

"Workstudy," web page, U.S. Department of Veterans Affairs, n.d. As of August 11, 2015:
http://www.benefits.va.gov/gibill/workstudy.asp

"Yellow Ribbon Reintegration Program," web page, Joint Services Support, n.d. As of August 11, 2015:
https://www.jointservicessupport.org/yrrp/default.aspx

Young, Ronald, director, Family and Employer Program and Policy, U.S. Department of Defense, testimony to United States Congress, House Committee on Veterans' Affairs, March 14, 2013.